Communications in Computer and Information Science   **708**

Commenced Publication in 2007
Founding and Former Series Editors:
Alfredo Cuzzocrea, Dominik Ślęzak, and Xiaokang Yang

More information about this series at http://www.springer.com/series/7899

Federico Rossi · Stefano Piotto
Simona Concilio (Eds.)

Advances in Artificial Life, Evolutionary Computation, and Systems Chemistry

11th Italian Workshop, WIVACE 2016
Fisciano, Italy, October 4–6, 2016
Revised Selected Papers

 Springer

Editors
Federico Rossi 🆔
Chemistry and Biology
University of Salerno
Fisciano
Italy

Simona Concilio 🆔
Department of Industrial Engineering
University of Salerno
Fisciano
Italy

Stefano Piotto 🆔
Department of Pharmacy
University of Salerno
Fisciano
Italy

ISSN 1865-0929   ISSN 1865-0937   (electronic)
Communications in Computer and Information Science
ISBN 978-3-319-57710-4   ISBN 978-3-319-57711-1   (eBook)
DOI 10.1007/978-3-319-57711-1

Library of Congress Control Number: 2017938634

Printed on acid-free paper

This Springer imprint is published by Springer Nature
The registered company is Springer International Publishing AG
The registered company address is: Gewerbestrasse 11, 6330 Cham, Switzerland

Preface

This volume of the Springer book series *Communications in Computer and Information Science* contains the proceedings of WIVACE 2016: the 11th Italian Workshop on Artificial Life and Evolutionary Computation, held in Salerno, Italy, during October 4–6, 2016. WIVACE was first held in 2007 in Sampieri (Ragusa), as the incorporation of two previously separately running workshops (WIVA and GSICE). After the success of the first edition, the workshop has been organized every year, aiming to offer a forum where different disciplines can effectively meet. The spirit of this workshop is to promote the communication among single research "niches" hopefully leading to surprising "cross-over" and "spill-over" effects. In this respect, the WIVACE community has been open to researchers coming from experimental fields such as systems chemistry and biology, origin of life, and chemical and biological smart networks.

WIVACE 2016 was jointly organized with BIONAM 2016, a workshop on bio-nanomaterials, to involve multidisciplinary research focusing on the analysis, synthesis and design, of bionanomaterials. The community of BIONAM comprises biophysicists, the biochemists, and bioengineers covering the study of the basic properties of materials and their interaction with biological systems, the development of new devices for medical purposes such as implantable systems, and new algorithms and methods for modeling the mechanical, physical, or biological properties of biomaterials. This challenging task requires powerful theoretical and computational tools to understand and control the inherent complexity of the interactions between synthetic and biological objects.

The interaction between the WIVACE and the BIONAM communities resulted in a joint session where the experimental work was harmonized in a well-established theoretical framework; some selected contributions, having a more theoretical character, have been collected in the section "Modelling and Simulation of Artificial and Biological Systems" of this volume.

The WIVACE 2016 volume is divided into two more sections: "Evolutionary Computation and Genetic Algorithms," which collects selected theoretical and computational contributions classically belonging to the WIVACE community, and "Systems Chemistry and Biology," which collects selected contributions from the interaction between informatics scientists and the biological and chemical community involved in complex systems studies. Among others, we would like to mention the contributions of two invited speakers, representative of this interaction: "Mathematical Modeling in Systems Biology" by Olli Yli-Harja and "A Strategy to Face Complexity: The Development of Chemical Artificial Intelligence" by Pier Luigi Gentili.

Events like WIVACE are generally a good opportunity for new-generation or soon-to-be scientists to get in touch with new subjects and bring new ideas to the attention of senior researchers. To highlight and promote the work of the youngest participants, we awarded ex aequo Dr. Chiara Damiani and Dr. Marcello Budroni for the best oral presentation; their contributions were selected as full papers and appear in this volume in the sections "Modelling and Simulation of Artificial and Biological

Systems" (C. Damiani et al.: "Linking Alterations in Metabolic Fluxes with Shifts in Metabolite Levels by Means of Kinetic Modeling") and "Evolutionary Computation and Genetic Algorithms" (M. Budroni et al.: "Scale-Free Networks out of Multifractal Chaos").

As editors, we wish to express gratitude to all the attendees of the conference and to the authors who spent time and effort to contribute to this volume. We also acknowledge the precious work of the reviewers and of the members of the Program Committee. Special thanks, finally, to the invited speakers for their very interesting and inspiring talks: Gabor Vattay from Eötvös Loránd University (Hungary), Nicola Segata from the University of Trento (Italy), Raffaele Giancarlo from the University of Palermo (Italy), Olli Yli-Harja from Tampere University of Technology (Finland), and Pier Luigi Gentili from University of Perugia (Italy).

The 17 papers presented were thoroughly reviewed and selected from 54 submissions. They cover the following topics: evolutionary computation, bioinspired algorithms, genetic algorithms, bioinformatics and computational biology, modelling and simulation of artificial and biological systems, complex systems, synthetic and systems biology, systems chemistry, and they represent the most interesting contributions to the 2016 edition of WIVACE.

October 2016

Federico Rossi
Stefano Piotto
Simona Concilio

Organization

WIVACE 2016 was organized in Fisciano (SA, Italy) by the University of Salerno (Italy).

Chairs

Federico Rossi	University of Salerno, Italy
Stefano Piotto	University of Salerno, Italy
Simona Concilio	University of Salerno, Italy

Program Committee

Amoretti Michele	University of Parma, Italy
Ballerini Lucia	University of Edinburgh, UK
Barba Anna Angela	University of Salerno, Italy
Bevilacqua Vitoantonio	Politecnico di Bari, Italy
Bocchi Leonardo	University of Florence, Italy
Cagnoni Stefano	University of Parma, Italy
Caivano Danilo	University of Bari, Italy
Cangelosi Angelo	University of Plymouth, UK
Carletti Timoteo	University of Namur, Belgium
Cattaneo Giuseppe	University of Salerno, Italy
Chella Antonio	University of Palermo, Italy
Concilio Simona	University of Salerno, Italy
Damiani Chiara	University of Milano-Bicocca, Italy
Favia Pietro	University of Bari, Italy
Filisetti Alessandro	Explora Biotech Srl, Italy
Fontanella Francesco	University of Cassino, Italy
Giacobini Mario	University of Turin, Italy
Graudenzi Alex	University of Milano-Bicocca, Italy
Marangoni Roberto	University of Pisa, Italy
Mauri Giancarlo	University of Milano-Bicocca, Italy
Mavelli Fabio	University of Bari, Italy
Moraglio Alberto	University of Exeter, UK
Nicosia Giuseppe	University of Catania, Italy
Nolfi Stefano	ISTC-CNR, Italy
Palazzo Gerardo	University of Bari, Italy
Pantani Roberto	University of Salerno, Italy
Piccinno Antonio	University of Bari, Italy
Piotto Stefano	University of Salerno, Italy
Pizzuti Clara	CNR-ICAR, Italy

Reverchon Ernesto	University of Salerno, Italy
Roli Andrea	University of Bologna, Italy
Rossi Federico	University of Salerno, Italy
Serra Roberto	University of Modena and Reggio, Italy
Spezzano Giandomenico	ICAR-CNR, Italy
Stano Pasquale	Roma Tre University, Italy
Terna Pietro	University of Turin, Italy
Tettamanzi Andrea	University of Nice Sophia Antipolis, France
Villani Marco	University of Modena and Reggio, Italy

Supported By

Società Chimica Italiana

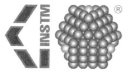

Contents

Evolutionary Computation, Genetic Algorithms and Applications

Scale-Free Networks Out of Multifractal Chaos. 3
  Marcello A. Budroni and Romualdo Pastor-Satorras

GPU-Based Parallel Search of Relevant Variable Sets in Complex Systems . . . 14
  Emilio Vicari, Michele Amoretti, Laura Sani, Monica Mordonini,
  Riccardo Pecori, Andrea Roli, Marco Villani, Stefano Cagnoni,
  and Roberto Serra

Complexity Science for Sustainable Smart Water Grids 26
  Angelo Facchini, Antonio Scala, Nicola Lattanzi, Guido Caldarelli,
  Giovanni Liberatore, Lorenzo Dal Maso, and Armando Di Nardo

New Paths for the Application of DCI in Social Sciences: Theoretical
Issues Regarding an Empirical Analysis. 42
  Riccardo Righi, Andrea Roli, Margherita Russo, Roberto Serra,
  and Marco Villani

MapReduce in Computational Biology - A Synopsis 53
  Giuseppe Cattaneo, Raffaele Giancarlo, Stefano Piotto,
  Umberto Ferraro Petrillo, Gianluca Roscigno, and Luigi Di Biasi

Photogrammetric Meshes and 3D Points Cloud Reconstruction:
A Genetic Algorithm Optimization Procedure. 65
  Vitoantonio Bevilacqua, Gianpaolo Francesco Trotta, Antonio Brunetti,
  Giuseppe Buonamassa, Martino Bruni, Giancarlo Delfine,
  Marco Riezzo, Michele Amodio, Giuseppe Bellantuono,
  Domenico Magaletti, Luca Verrino, and Andrea Guerriero

Benchmarking Spark Distributed Data Structures: A Sequence
Analysis Case Study 77
  Umberto Ferraro Petrillo and Roberto Vitali

Modelling and Simulation of Artificial and Biological Systems

Automatic Design of Boolean Networks for Cell Differentiation 91
  Michele Braccini, Andrea Roli, Marco Villani, and Roberto Serra

Model-Based Lead Molecule Design 103
  Alessandro Giovannelli, Debora Slanzi, Marina Khoroshiltseva,
  and Irene Poli

Reducing Dimensionality in Molecular Systems: A Bayesian
Non-parametric Approach 114
   Valentina Mameli, Nicola Lunardon, Marina Khoroshiltseva,
   Debora Slanzi, and Irene Poli

Constraint-Based Modeling and Simulation of Cell Populations 126
   Marzia Di Filippo, Chiara Damiani, Riccardo Colombo, Dario Pescini,
   and Giancarlo Mauri

Linking Alterations in Metabolic Fluxes with Shifts in Metabolite Levels
by Means of Kinetic Modeling 138
   Chiara Damiani, Riccardo Colombo, Marzia Di Filippo, Dario Pescini,
   and Giancarlo Mauri

Systems Chemistry and Biology

A Strategy to Face Complexity: The Development of Chemical
Artificial Intelligence..................................... 151
   Pier Luigi Gentili

Mathematical Modeling in Systems Biology....................... 161
   Olli Yli-Harja, Frank Emmert-Streib, and Jari Yli-Hietanen

Synchronization in Near-Membrane Reaction Models of Protocells 167
   Giordano Calvanese, Marco Villani, and Roberto Serra

On the Employ of Time Series in the Numerical Treatment
of Differential Equations Modeling Oscillatory Phenomena 179
   Raffaele D'Ambrosio, Martina Moccaldi, Beatrice Paternoster,
   and Federico Rossi

A Program for the Solution of Chemical Equilibria Among
Multiple Phases 188
   Fulvio Ciriaco, Massimo Trotta, and Francesco Milano

Author Index 199

Evolutionary Computation, Genetic Algorithms and Applications

Scale-Free Networks Out of Multifractal Chaos

Marcello A. Budroni[1(✉)] and Romualdo Pastor-Satorras[2]

[1] Nonlinear Physical Chemistry Unit,
Service de Chimie Physique et Biologie Théorique, Université libre de Bruxelles,
CP 231 - Campus Plaine, 1050 Brussels, Belgium
mbudroni@ulb.ac.be, mabudroni@uniss.it
[2] Departament de Física, Universitat Politècnica de Catalunya,
Campus Nord B4, 08034 Barcelona, Spain
romualdo.pastor@upc.edu
http://physchem.uniss.it/cnl.dyn/budroni.html

Abstract. Fractal and multifractal properties characterize many real-world scale-free networks. Here we present a deterministic approach to generate power-law networks from multifractal chaotic time series. We show, both analytically and numerically, how the resulting scale-free topologies preserve the multifractal information of the original chaotic source embedded in the exponent of the power-law degree distribution.

Keywords: Multifractal processes · Power-law networks · Chaotic dynamics

1 Introduction

Understanding complex and aperiodic phenomena encountered in biology [27], chemistry [10,16,25,32], economics [7] and physics [6,9,17], represents an open scientific challenge. The progress towards this fundamental goal can benefit from different theoretical frameworks, including statistical physics and complex network theory, information theory, non-linear dynamics and chaos, that constitute the composite panorama of Complex Science. In this context any effort to find synergies among different approaches greatly helps to move steps forward in controlling complexity. Our contribution here is concerned at presenting a possible pathway to relate chaos and network theory.

During the last years, complex network theory has rapidly grown as a interpretative framework for many complex systems and phenomena, ranging from financial crises to epidemics spreading [6]. Though this approach may appear as a drastic simplification of the specific features of a system constituents, it is able to disentangle the intrinsic topology of their interactions, which crucially impacts the possible dynamics running on the network itself [31].

In the realm of dynamical systems, network statistical techniques have been applied to analyse nonlinear time series, with a particular focus on characterizing chaotic dynamics. The main idea of this methodology is to transform the information of a time series from the temporal domain into the topology

© Springer International Publishing AG 2017
F. Rossi et al. (Eds.): WIVACE 2016, CCIS 708, pp. 3–13, 2017.
DOI: 10.1007/978-3-319-57711-1_1

of a network and, hence, the key point resides in the way one defines nodes and links. So far, several transformation approaches have been proposed [2, 11–14, 19, 20, 24, 26, 28, 33, 35–38] and a bench of network tools have been adapted to the analysis of nonlinear time series.

However, less effort has been devoted to investigate how the latter could, in turn, be exploited as a source for growing complex network with non-trivial connectivity patterns. Most of real-world networks are inhomogeneous, showing scale-free property defined by a power-law degree distribution $P(k) \sim k^{-\gamma}$, where k is the number of connections of a node (degree). This feature has been successfully explained through preferential attachment mechanisms [5]. In these mechanisms nodes that stochastically gain a higher degree, present also stronger ability to attract new links added to the network, leading to the formation of structures with a small number of highly connected nodes in spite of a broad spectrum of moderately and scarcely connected nodes.

Recently, it has been pointed out how an intrinsic aspect of this hierarchical connectivity is the presence of fractal and self-similar features embedded in the network topology. Stimulated by the seminal paper by Song et al. [34], fractal properties of scale-free networks have been revealed and measured by adapting box-counting approaches to the non-euclidean geometry of complex networks. In particular, networks were suitably partitioned into sub-graphs or clusters with characteristic diameters (in the sense of network distance) and self-similarity was shown when scaling this characteristic measure. Following similar *a posteriori* partition strategies, the possibility for multifractality has been also analytically demonstrated by Furuya and Yakubo [18] and attributed to the large fluctuations of local node density in scale-free networks.

In this context, an open question is whether (and which) deterministic multifractal processes could be considered *a priori* as alternative evolution mechanisms for growing scale-free networks that preserve the multifractality of the original source in the ultimate structure.

In this paper we present a novel model for developing power-law networks starting from a multifractal chaotic generator of numbers. We show that the resulting topologies preserve the multifractal nature of the underlying chaotic source and we also derive analytically the relation which ties the power-law exponent characterizing the connectivity of these networks with the generalized dimension of the projected dynamics. Finally, we discuss this closed-form relation as a stable tool for characterizing the multifractal spectrum of a time series through the analysis of the network connectivity.

2   Model

We generate networks from chaotic dynamical data by means of a transition transformation introduced in [11] and briefly resumed hereunder. We start with the set $\mathcal{V} = \{M \text{ nodes}\}$ and the network connectivity is built-up by using a normalized chaotic series of numbers $\mathcal{G}_{chaotic} = \{x_j : x_j \in \mathbb{R} : [0, 1], j \in [1, n]\}$, where $n >> 1$ is the size of $\mathcal{G}_{chaotic}$. Nodes are identified with the index

$i = \lfloor x_j M + 1 \rfloor$ (where $\lfloor z \rfloor$ is the floor function) and an undirected connection between two successive nodes $i = \lfloor x_j M + 1 \rfloor$ and $l = \lfloor x_{j+1} M + 1 \rfloor$ $(i, l \in \mathcal{V})$ is established if it does not constitute any multiple–connection. When these criteria are not met, the successive pair of numbers, namely $i' = \lfloor x_{j+1} M + 1 \rfloor$ and $l' = \lfloor x_{j+2} M + 1 \rfloor$, is considered. The previous step is reiterated until the maximal possible number of edges is introduced in the network, i.e. until a stationary network is achieved.

The structures resulting from this procedure are connected networks by construction, preserve temporal information of the generator and, because of the peculiar fractal properties of the strange attractors underlying chaotic sources, consist of a fraction $N(M)$ of the initial M nodes. In this framework, the network provides an alternative way for partitioning the fractal support of the chaotic dynamics congruent with the box-counting method [1,21,22], where the $N(M)$ nodes of the network correspond to the number of boxes of length $\epsilon = M^{-1}$ needed to cover the fractal chaotic attractor in the phase space. As a consequence, the maximal number of edges asymptotizes to the upper limit, $\mathcal{L}_{chaotic}(M)$, which is characteristic of the chaotic source at hand and is strictly lower than the fully connected configuration $M(M-1)/2$. $N(M)$ and $\mathcal{L}_{chaotic}(M)$ are related to the fractal dimension of the chaotic series as [11]:

$$\mathcal{L}_{chaotic}(M) \approx \frac{\langle k \rangle}{2} N(M) \frown M^{D_0}, \tag{1}$$

where D_0 is the capacity dimension of the set (obtained through the linear regression of $\log(N(M))$ versus $\log(M)$) and $\langle k \rangle$ is the average degree of the network. In our previous work [11] $\mathcal{L}_{chaotic}(M)$ was used as a topological observable for (i) characterizing the capacity dimension of a chaotic series and (ii) discerning chaotic dynamics from random ones, being the latter capable of realizing fully connected configurations.

In this work we want to study more in detail the connectivity (typically the degree distribution) of the these networks and relate them to the multifractality of the underlying chaotic attractor. To do so, we consider a paradigmatic example of chaotic generators, the logistic map $x_{j+1} = r\, x_j (1 - x_j)$. This discrete-time formula maps the interval $x \in [0, 1]$ into itself when the control parameter r ranges between 0 and 4. Multifractal chaotic regimes interspersed with periodic windows occur in the interval $r \in [3.57, 4)$ and hereunder we will consider the representative case $r = 3.7$ to back up the validity of the following analytical approach. The map is iterated as needed to achieve a stationary connectivity in the network (typically $n \sim 10^3 M$). In this sense possible finite-size effects of the chaotic time series are ruled out.

3   Scale-Free Networks Out of Chaos

When the algorithm described above is applied to the multifractal logistic source, the emerging networks exhibit characteristic scale-free properties as indicated by a power-law degree distribution with an exponent around 3. In Fig. 1 we report the cumulative degree distribution $P_{cum}(k) = \frac{1}{N(M)} \sum_{i/k_i \leq k} \mathbb{1}$ (giving the probability that a network node presents degree equal or larger than k) of the logistic network. The plot describes the scale-free nature of networks for different sizes ($M \in [10^4, 10^7]$) with all trends collapsing to a common power-law distribution $P_{cum}(k) = k^{-\gamma'}$ characterized by $\gamma' \sim 2.142(3)$. The exponent of the simple degree distribution $P(k)$ then reads $\gamma = \gamma' + 1 \sim 3.142(3)$. Power-law scale-invariant properties have been obtained for networks generated from other values of the critical parameter r of the logistic map (in the range where it presents multifractal characteristics) and from other 1-dimensional maps [29].

In the following analysis we prove that this power-law trends in the degree distribution reflect the multifractal nature of the network and can be analytically related to the generalized dimension of the chaotic generator.

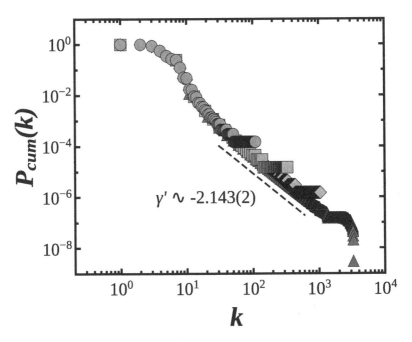

Fig. 1. Cumulative degree distributions of the logistic network ($r = 3.7$) for $M = 10^4$ (red circles), 10^5 (green squares), 10^6 (grey diamonds) and 10^7 (blue triangles) nodes. $n = 1 \times 10^{10}$ iterations and $P_{cum}(k)$ is averaged over 100 networks (i.e. 100 different initial seeds of the chaotic generator). (Color figure online)

For strange attractors it is common that different regions are differently visited, and chaotic orbits will spend most of their time in a small minority of the $N(\epsilon)$ boxes partitioning the fractal support underneath the chaotic attractor itself. An illustration of this property is given in Fig. 2a for the unidimensional support of the logistic map with $r = 3.7$. The dimension D_q takes into account these heterogeneous probability pattern and generalizes the definition of the box-counting dimension as

$$D_q = \frac{1}{q-1} \lim_{\epsilon \to 0} \frac{\log \sum_i^{N(\epsilon)} p_i^q}{\log \epsilon}. \tag{2}$$

This characterizes the intrinsic hierarchy within a fractal set in terms of the moments q of the partition function $\sum_i^{N(\epsilon)} p_i^q$ [22,29,30]. Here $p_i = \lim_{n \to \infty} \frac{n_i}{n}$ quantifies the probability, termed *natural measure*, that the chaotic map returns in the i-th box of the $N(\epsilon)$ available boxes, during an infinitely long orbit (in practise n_i times over $n \gg 1$ iterations of the chaotic orbit). $D_q(q)$ exhibits a non-constant scaling bounded between the asymptotic values $D_{\pm\infty}$ when a heterogeneous probability distribution describes the recurrence of a chaotic trajectory over different regions of the attractor which can thus be defined multifractal. An example of such a case is shown in Fig. 2b, where we report the cumulative distribution of the natural measure, $P_{cum}(p)$, for the logistic map displayed in Fig. 2a. It can be observed how this trend describes an extremely heterogeneous statistics and, in particular, follows a power-law behaviour (Fig. 2b), characterized by the same exponent as for the cumulative degree distribution of the associated graph (compare Figs. 1 and 2b).

From this evidence stems the initial ansatz of our analytical approach, where we assume that the degree of network nodes is representative of the natural measure of the corresponding boxes partitioning the fractal support. In particular, as a first approach, we can reasonably hypothesize that an increasing linear relation links the degree k of a certain node to the natural measure p of the associated box.

Thanks to this correlation, we can re-write the natural measures involved in the computation of D_q in terms of node degrees through

$$p_i \sim \frac{k_i}{\langle k \rangle N(\epsilon)}. \tag{3}$$

Since in scale-free networks the average degree $\langle k \rangle$ is a constant [6,15] and can be neglected in relation (3), the partition sum of Eq. (2) reads

$$\sum_i p_i^q \sim \sum_i \left(\frac{k_i}{N(\epsilon)} \right)^q \sim N(\epsilon)^{-q} \sum_i \frac{k_i^q}{N(\epsilon)} N(\epsilon) \tag{4}$$

$$\sim \langle k^q \rangle N(\epsilon)^{1-q}.$$

From the definition of D_q (see Eq. (2)), it follows

$$\sum_i p_i^q \sim \epsilon^{(q-1)D_q} \sim N(\epsilon)^{-(q-1)D_q/D_0} \tag{5}$$

(being $\epsilon = N(\epsilon)^{-(1/D_0)}$). By comparing Eqs. (4) and (5), we find that

$$\langle k^q \rangle N(\epsilon)^{1-q} \sim N(\epsilon)^{-(q-1)D_q/D_0} \tag{6}$$

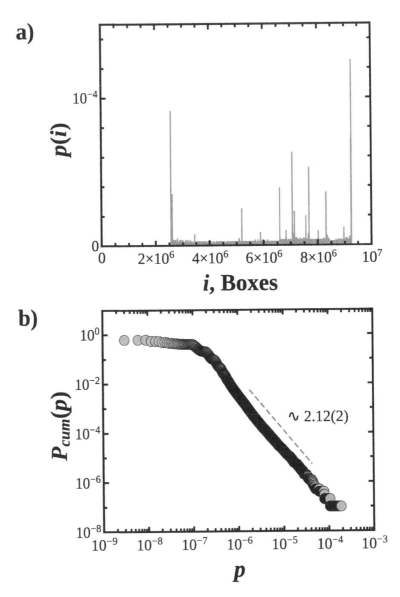

Fig. 2. (a) Natural measure $p(i)$ of the i-th box for the logistic map with $r = 3.7$. The support $[0, 1]$ is partitioned in $M = 1 \times 10^7$ boxes; the map is iterated for $n = 1 \times 10^{10}$ time steps and the statistics is performed over 100 initial conditions. (b) Cumulative probability distribution, $P_{cum}(p)$, of the box natural measures $p(i)$ for the logistic map with $r = 3.7$, $M = 1 \times 10^7$ boxes, $n = 1 \times 10^{10}$ iterations. $P_{cum}(p)$ is averaged over 100 initial conditions.

and, hence

$$\langle k^q \rangle \backsim N(\epsilon)^{(q-1)(1-D_q/D_0)}. \tag{7}$$

which features a first expression relating a topological observable of the network and the generalized dimension of the multifractal chaotic source.

$\langle k^q \rangle$ can be also written as

$$\langle k^q \rangle = \int_{m(\epsilon)}^{k_c(\epsilon)} dk \, P(k) \, k^q \tag{8}$$

where $P(k) \backsim k^{-\gamma+1}$ is the degree distribution of the projected network, $[m(\epsilon), k_c(\epsilon)]$ is the k–domain where $P(k)$ exhibits the power-law tail and the degree cut-off $k_c(\epsilon)$ is the maximal degree of the network. In scale-free networks, when the exponent of the integral argument is $q - \gamma > 0$, integral (8) is asymptotically equal to $k_c(\epsilon)^{q+1-\gamma}$ [6,15] and quickly diverges as the size of the network tends to the thermodynamic limit. Keeping in mind Eq. (7), it can then be shown that for large positive values $q = q_\infty$ where D_q saturates to D_∞, integral (8) reads

$$k_c(\epsilon)^{q_\infty} \backsim N(\epsilon)^{q_\infty (1-D_\infty/D_0)}, \tag{9}$$

implying

$$k_c(\epsilon) \backsim N(\epsilon)^{(1-D_\infty/D_0)}. \tag{10}$$

Equation (10) ties $k_c(\epsilon)$ and D_∞ through D_0. In detail, D_∞ can be extrapolated from the slope $\beta = 1 - D_\infty/D_0$ of the linear regression of $\log(k_c(\epsilon))$ plotted versus $\log(N(\epsilon))$ (see Fig. 3) following

$$D_\infty = D_0(1 - \beta), \tag{11}$$

where D_0 is known from Eq. (1). The second relation for $\langle k^q \rangle$ is thus derived by substituting $k_c(\epsilon)$ in Eq. (8), to obtain

$$\langle k^q \rangle \backsim N(\epsilon)^{(q+1-\gamma)(1-D_\infty/D_0)}. \tag{12}$$

Finally, combining Eqs. (7) and (12)

$$(q - 1)(1 - D_q/D_0) = (q + 1 - \gamma)(1 - D_\infty/D_0) \tag{13}$$

and, conveniently re–arranging,

$$\gamma = 2 + (q - 1)\frac{D_q - D_\infty}{D_0 - D_\infty}. \tag{14}$$

one can laid down a closed-form relating γ to D_0, D_q and D_∞. This expresses the "latent" multifractality of a scale-free network grown from the projection of a multifractal chaotic series and describes how multifractal measures are quantitatively incorporated in the power-law exponent.

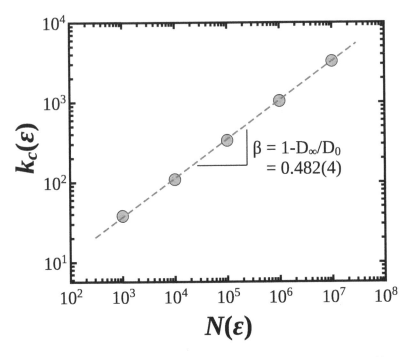

Fig. 3. Scaling of the degree cut-off, $k_c(\epsilon)$, as function of the network size $N(\epsilon)$. D_0 and D_∞ can be computed by means of Eqs. (1) and (11), respectively. For this illustrative case $\beta = 0.482 \pm 0.004$ and $D_0 \sim 1$.

4   Concluding Discussion

In this paper we have presented a new perspective to bridge chaotic dynamics and complex networks. Specifically, we have introduced a simple procedure able to grow scale-free networks by using generators of multifractal chaotic series. The heterogeneity and the free–scale nature of these networks, encoded in the power-law degree distribution $P(k)$, are demonstrated to be analytically related to the multifractal properties of the generating chaotic source. While fractal and multifractal properties of many real scale-free networks have been already unveiled through *a posteriori* analysis, our model shows that a chaotic multi-fractal processes can represent an *a priori* mechanism for growing power-law networks which, in turn, preserve multifractal information of the original source in the ultimate topology. With respect to the stochastic preferential attachment mechanisms chaotic generators could be seen as an alternative deterministic pathway for the formation of scale-free structures.

In our numerical exploration we found that a multifractal process can poten-tially be mapped into a power-law network if (*i*) a linear relation ties the natural measures to the degrees of the nodes and (*ii*) the distribution of the natural measures shows a power-law trend. Work is in progress [8] to generalize this description to cases in which the natural measure increases nonlinearly with the node degree. From a network analysis viewpoint other topological properties, such as clustering and assortativity within these multifractal networks should be investigated in depth in order to unravel further correlations between the network connectivity and the properties of underlying chaotic dynamics.

From the perspective of time series analysis, this work represents a further proof of concept of the great potential of network approaches when applied to the characterization of nonlinear dynamics. Thanks to a simple statistics on the network connectivity it is possible to calculate the generalized dimension of the associated chaotic generator *via* a closed formula. This can be exploited as a robust method for multifractal analysis, particularly stable for high indexes q of the generalized dimension, prohibitive to box-counting methods.

The validity of our approach is demonstrated here for the theoretical but still general study case of 1-dimensional logistic-like maps. A future domain of investigation is the case of multifractal series resulting from non-chaotic processes, like binomial multifractal generators [23]. Also our challenge is to extend this framework to real multifractal normalized time series of practical interest. Prominent examples are time series collecting earthquakes frequency and magnitude, that have been proven to converge into universal power-law descriptions [4]. In this context fractal and multifractal measures are of utmost interest and network theory is already fruitfully applied to disclose the highly hierarchical and complex spatio-temporal organization of these phenomena and improve predictive protocols [3].

Acknowledgments. The authors thank A. Baronchelli for fruitful discussions. M.A.B. is supported by FRS-FNRS. R.P.-S. acknowledges financial support from the Spanish MINECO, under project FIS2013-47282-C2-2, and EC FET-Proactive Project MULTIPLEX (Grant No. 317532) and from ICREA Academia, funded by the Generalitat de Catalunya.

References

1. Grassberger, P.: On efficient box-counting algorithms. Int. J. Mod. Phys. C **4**(3), 515–523 (1993)
2. Yang, Y., Yang, H.: Complex network-based time series analysis. Physica A: Stat. Mech. Appl. **387**(5–6), 1381–1386 (2008)
3. Abe, S., Pastén, D., Suzuki, N.: Finite data-size scaling of clustering in earthquake networks. Physica A: Stat. Mech. Appl. **390**(7), 1343–1349 (2011). http://www.sciencedirect.com/science/article/pii/S0378437110009970
4. Bak, P., Christensen, K., Danon, L., Scanlon, T.: Unified scaling law for earthquakes. Phys. Rev. Lett. **88**, 178501 (2002). http://link.aps.org/doi/10.1103/PhysRevLett.88.178501
5. Barabási, A.L., Albert, R.: Emergence of scaling in random networks. Science **286**(5439), 509–512 (1999)
6. Barrat, A., Barthelemy, M., Vespignani, A.: Dynamical Processes on Complex Networks. Cambridge University Press, Cambridge (2008)
7. Brian Arthur, W.: Complexity and the Economy. Oxford Economic Press, Oxford (1997)

8. Budroni, M.A., Baronchelli, A., Pastor-Satorras, R.: Scale-free networks emerging from multifractal time series. ArXiv e-prints, December 2016
9. Budroni, M.A., Lemaigre, L., De Wit, A., Rossi, F.: Cross-diffusion-induced convective patterns in microemulsion systems. Phys. Chem. Chem. Phys. **17**, 1593–1600 (2015). http://dx.doi.org/10.1039/C4CP02196G
10. Budroni, M.A., Pilosu, V., Delogu, F., Rustici, M.: Multifractal properties of ball milling dynamics. Chaos Interdisc. J. Nonlinear Sci. **24**(2), 023117 (2014). http://dx.doi.org/10.1063/1.4875259
11. Budroni, M.A., Tiezzi, E., Rustici, M.: On chaotic graphs: a different approach for characterizing aperiodic dynamics. Physica A Stat. Mech. Appl. **389**(18), 3883–3891 (2010). http://www.sciencedirect.com/science/article/pii/S0378437110004796
12. Campanharo, A.S., Irmak Sirer, M., Dean Malmgren, R., Ramos, F.M., Nunes Amaral, L.A.: Duality between time series and networks. Plos One **6**(8), e23378 (2011)
13. Donner, R.V., Small, M., Donges, J.F., Marwan, N., Zou, Y., Xiang, R., Kurths, J.: Recurrence-based time series analysis by means of complex network methods. Int. J. Bifurcat. Chaos **21**, 1019–1046 (2011)
14. Donner, R.V., Zou, Y., Donges, J.F., Marwan, N., Kurths, J.: Recurrence networks: a novel paradigm for nonlinear time series analysis. New J. Phys. **12**(3), 033025 (2010)
15. Dorogovtsev, S.N., Mendes, J.F.F.: Evolution of networks. Adv. Phys. **51**(4), 1079–1187 (2002)
16. Escala, D.M., Budroni, M.A., Carballido-Landeira, J., De Wit, A., Muñuzuri, A.P.: Self-organized traveling chemo-hydrodynamic fingers triggered by a chemical oscillator. J. Phys. Chem. Lett. **5**(3), 413–418 (2014)
17. Facchini, A., Wimberger, S., Tomadin, A.: Multifractal fluctuations in the survival probability of an open quantum system. Physica A: Stat. Mech. Appl. **376**, 266–274 (2007). http://www.sciencedirect.com/science/article/pii/S0378437106010582
18. Furuya, S., Yakubo, K.: Multifractality of complex networks. Phys. Rev. E **84**, 036118 (2011). http://link.aps.org/doi/10.1103/PhysRevE.84.036118
19. Gao, Z., Jin, N.: Complex network from time series based on phase space reconstruction. Chaos **19**(3), 033137 (2009)
20. Gao, Z., Jin, N.: Flow-pattern identification and nonlinear dynamics of gas-liquid two-phase flow in complex networks. Phys. Rev. E **79**, 066303 (2009)
21. Grassberger, P., Procaccia, I.: Measuring the strangeness of strange attractors. Physica D: Nonlinear Phenom. **9**(1), 189–208 (1983). http://www.sciencedirect.com/science/article/pii/0167278983902981
22. Halsey, T.C., Jensen, M.H., Kadanoff, L.P., Procaccia, I., Shraiman, B.I.: Fractal measures and their singularities: the characterization of strange sets. Phys. Rev. A **33**, 1141–1151 (1986). http://link.aps.org/doi/10.1103/PhysRevA.33.1141
23. Kantelhardt, J.W., Zschiegner, S.A., Koscielny-Bunde, E., Havlin, S., Bunde, A., Stanley, H.: Multifractal detrended fluctuation analysis of nonstationary time series. Physica A: Stat. Mech. Appl. **316**(1–4), 87–114 (2002). http://www.sciencedirect.com/science/article/pii/S0378437102013833
24. Lacasa, L., Luque, B., Ballesteros, F., Luque, J., Nuño, J.C.: From time series to complex networks: the visibility graph. Proc. Natl. Acad. Sci. U.S.A. **105**, 4972–4975 (2008)

25. Marchettini, N., Budroni, M.A., Rossi, F., Masia, M., Turco Liveri, M.L., Rustici, M.: Role of the reagents consumption in the chaotic dynamics of the Belousov-Zhabotinsky oscillator in closed unstirred reactors. Phys. Chem. Chem. Phys. **12**, 11062–11069 (2010)

26. Marwan, N., Donges, J.F., Zou, Y., Donner, R.V., Kurths, J.: Complex network approach for recurrence analysis of time series. Phys. Lett. A **373**(46), 4246–4254 (2009). http://www.sciencedirect.com/science/article/pii/S0375960109011852

27. Murray, J.D.: Mathematical Biology. Springer, New York, USA (2002)

28. Nicolis, G., Garcia Cantu, A., Nicolis, C.: Dynamical aspects of interaction networks. Int. J. Bifurcat. Chaos **15**(11), 3467 (2005)

29. Ott, E.: Chaos in Dynamical Systems. Cambridge University Press, Cambridge (1993)

30. Paladin, G., Vulpiani, A.: Anomalous scaling laws in multifractal objects. Phys. Rep. **156**(4), 147–225 (1987). http://www.sciencedirect.com/science/article/pii/0370157387901104

31. Pastor-Satorras, R., Vespignani, A.: Epidemic spreading in scale-free networks. Phys. Rev. Lett. **86**, 3200–3203 (2001). http://link.aps.org/doi/10.1103/PhysRevLett.86.3200

32. Rossi, F., Budroni, M.A., Marchettini, N., Cutietta, L., Rustici, M., Turco Liveri, M.L.: Chaotic dynamics in an unstirred ferroin catalyzed Belousov-Zhabotinsky reaction. Chem. Phys. Lett. **480**(4–6), 322–326 (2009). http://www.sciencedirect.com/science/article/pii/S0009261409011087

33. Shirazi, A.H., Reza Jafari, G., Davoudi, J., Peinke, J., Reza Rahimi Tabar, M., Sahimi, M.: Mapping stochastic processes onto complex networks. J. Stat. Mech. **07**, P07046 (2009)

34. Song, C., Havlin, S., Makse, H.A.: Self-similarity of complex networks. Nature (Lond.) **433**(2), 392–395 (2005)

35. Sun, X., Small, M., Zhao, Y., Xue, X.: Characterizing system dynamics with a weighted and directed network constructed from time series data. Chaos **24**(2), 024402 (2014). http://scitation.aip.org/content/aip/journal/chaos/24/2/10.1063/1.4868261

36. Xiang, R., Zhang, J., Xu, X.K., Small, M.: Multiscale characterization of recurrence-based phase space networks constructed from time series. Chaos **22**(1), 013107 (2012)

37. Zhang, J., Small, M.: Complex network from pseudoperiodic time series: topology versus dynamics. Phys. Rev. Lett. **96**, 238701 (2006)

38. Zou, Y., Donner, R.V., Thiel, M., Kurths, J.: Disentangling regular and chaotic motion in the standard map using complex network analysis of recurrences in phase space. Chaos **26**(2), 023120 (2016)

GPU-Based Parallel Search of Relevant Variable Sets in Complex Systems

Emilio Vicari[1], Michele Amoretti[1], Laura Sani[1], Monica Mordonini[1],
Riccardo Pecori[1,4], Andrea Roli[2], Marco Villani[3], Stefano Cagnoni[1(✉)],
and Roberto Serra[3]

[1] Dipartimento di Ingegneria ed Architettura, Università di Parma, Parma, Italy
stefano.cagnoni@unipr.it
[2] Dip. di Informatica, Scienza e Ingegneria,
Università di Bologna - Sede di Cesena, Cesena, Italy
[3] Dip. Scienze Fisiche, Informatiche e Matematiche,
Università di Modena e Reggio Emilia, Modena, Italy
[4] SMARTest Research Centre, Università eCAMPUS, Novedrate, CO, Italy

Abstract. Various methods have been proposed to identify emergent dynamical structures in complex systems. In this paper, we focus on the Dynamical Cluster Index (DCI), a measure based on information theory which allows one to detect relevant sets, i.e. sets of variables that behave in a coherent and coordinated way while loosely interacting with the rest of the system. The method associates a score to each subset of system variables; therefore, for a thorough analysis of the system, it requires an exhaustive enumeration of all possible subsets. For large systems, the curse of dimensionality makes the problem solvable only using metaheuristics. Even within such approaches, however, DCI computation has to be performed for a huge number of times; thus, an efficient implementation becomes a mandatory requirement. Considering that a candidate relevant set's DCI can be computed independently of the others, we propose a GPU-based massively parallel implementation of DCI computation. We describe the algorithm's structure and validate it by assessing the speedup in comparison with a single-thread sequential CPU implementation when analyzing a set of dynamical systems of different sizes.

Keywords: GPU-based parallel programming · Complex systems · Relevant sets

1 Introduction

The behavior of a complex system can be described by identifying emergent dynamical structures within it, i.e., subsets of variables whose members tightly interact with (depend on) one another, as well as hierarchically, by identifying higher-level interactions that occur between such sets.

The study of complex systems is related to the identification of emergent properties of systems whose components are usually well-known and defined in

© Springer International Publishing AG 2017
F. Rossi et al. (Eds.): WIVACE 2016, CCIS 708, pp. 14–25, 2017.
DOI: 10.1007/978-3-319-57711-1_2

terms of state variables. To describe the organization of complex systems several measures of complexity have been proposed, many of which based on information theory (as, for instance, in [4,6]).

Many different systems can be described effectively in terms of coordinated dynamical behavior of groups of elements; for example, relevant examples in the domain of neuroscience can be found in [8,9].

Tononi et al. [10], and later other authors (Sporns et al. [9], Villani et al. [12]) introduced a method to identify relevant structures in complex systems. Based on a data-set including samples of the system status at different times, one can associate each possible subset of variables with an index T_c. Such an index quantifies how much its behavior deviates from the behavior of a reference (homogeneous) system, in which the variables have, individually, the same distribution as in the data-set, but are homogeneously correlated. Therefore, the higher its T_c, the higher the degree of correlation/interaction between the variables in a subset. The subsets characterized by high T_c values are referred to as Candidate Relevant Sets (CRSs), the properly called Relevant Subsets (RSs) being candidates that do not include (or are not included in) other candidate sets with higher T_c values [12].

For a complete description of the dynamical system, T_c must be computed for each possible set, which becomes unfeasible as the dimension of the system increases. Subsets of variables describing high-dimensional systems can therefore be identified by using a metaheuristic which smartly explores the search space [7]. Even in this case, T_c computation must be repeated hundreds of thousands to millions of times. An efficient implementation of such a function is therefore definitely necessary. Considering that the computation of T_c for each candidate RS is independent of the others, using GPU-based parallel code seems to be the most efficient way of computing the index.

We have developed a set of CUDA C[1] kernels that provide a fine-grained parallel implementation of the main building blocks needed to compute the T_c index, upon which smart and efficient search algorithms can be designed.

The parallel functions were developed to accomplish three different goals in our study:

1. Speeding up an exhaustive sequential search by computing the T_c values of several candidate RSs in parallel;
2. Providing a computationally-efficient objective function for a metaheuristic that searches for the RSs of large dynamical systems for which an exhaustive search is impractical;
3. Making it possible to explore more complex systems and detect possible hierarchical dependencies between RSs.

In the next section, we briefly introduce the basics of the method for which we have developed the CUDA kernels. Then we analyze the computational problem, identifying the algorithm blocks that are most amenable to parallelization, and describe their GPU-based implementation. We conclude our paper by reporting

---

[1] https://developer.nvidia.com.

the results of the tests in which we compare the performance of our parallel code with respect to a standard single-CPU sequential implementation. Finally, in the last section, we foresee possible future steps in our research that we expect the development of the parallel code to make feasible.

2   Method

In this section we succinctly illustrate the procedure for computing the T_c. The interested reader can find more details in [3,12].

Let the system under exam be modeled by means of a set U of N variables, which assume finite and discrete values. The *cluster index* of a subset S of variables in U, $S \subset U$, as defined by Tononi et al. [10], estimates the ratio between the amount of information integration among the variables in S and the amount of integration between S and U. These quantities depend on Shannon's entropy of both the single elements and the sets of elements in U.

The entropy of an element x_i is defined as:

$$H(x_i) = - \sum_{v \in V_i} p(v) \ log \ p(v) \tag{1}$$

where V_i is the set of the possible values of x_i and $p(v)$ the probability of occurrence of symbol v. The entropy of a pair of elements x_i and x_j is defined by means of their joint probabilities:

$$H(x_i, x_j) = - \sum_{v \in V_i} \sum_{w \in V_j} p(v, w) \ log \ p(v, w) \tag{2}$$

Equation 2 can be extended to sets of k elements considering the probability of occurrence of vectors of k values. This approach deals with observational data, therefore probabilities are estimated by means of relative frequencies.

The cluster index $C(S)$ of a set S of k elements is defined as the ratio between the integration $I(S)$ of S and the mutual information between S and the rest of the system $U - S$.

The integration of subset S is defined as:

$$I(S) = \sum_{x \in S} H(x) - H(S) \tag{3}$$

$I(S)$ represents the deviation from statistical independence of the k elements in S. The mutual information $M(S; U - S)$ is defined as:

$$M(S; U - S) \equiv H(S) + H(S|U - S) = H(S) + H(U - S) - H(S, U - S) \tag{4}$$

where $H(A|B)$ is the conditional entropy and $H(A, B)$ the joint entropy. Finally, the cluster index $C(S)$ is defined as:

$$C(S) = \frac{I(S)}{M(S; U - S)} \tag{5}$$

Since C is defined as a ratio, it is undefined in all those cases where $M(S; U - S)$ vanishes. In this case, the subset S is statistically independent from the rest of the system and needs to be analyzed separately. As $C(S)$ scales with the size of S, cluster index values of systems of different size need to be normalized. To this aim, a reference system is defined, i.e., the homogeneous system U_h, randomly generated according to the probability distribution of each state of the original system U. Then, for each subsystem size of U_h the average integration $\langle I_h \rangle$ and the average mutual information $\langle M_h \rangle$ are computed. Finally, the cluster index value of S is normalized by means of an appropriate normalization constant:

$$C'(S) = \frac{I(S)}{\langle I_h \rangle} / \frac{M(S; U - S)}{\langle M_h \rangle} \tag{6}$$

Furthermore, to assess the significance of the differences observed in the cluster index values, a statistical index T_c is computed:

$$T_c(S) = \frac{C'(S) - \langle C'_h \rangle}{\sigma(C'_h)} \tag{7}$$

where $\langle C'_h \rangle$ and $\sigma(C'_h)$ are the average and the standard deviation of the population of normalized cluster indices with the same size as S from the homogeneous system.

We emphasize that the indices in 5–7 are defined without any reference to a particular type of system. In their original papers, Edelman and Tononi considered the fluctuations of a neural system around a stationary state. In our approach, this measure is applied to time series of data generated by a dynamical model. In general, these data lack the stationary properties of fluctuations around a fixed point. Moreover, depending upon the case at hand, either transients from arbitrary initial states to a final attractor, or collections of attractor states can be considered, as well as responses to perturbations of attractor states. In all these cases we will use Eq. 5, that will therefore be called the Dynamical Cluster Index (DCI), as it aims at detecting subsets of variables that are relevant to the system's dynamics.

The search for relevant subsets of variables of a dynamical system by means of the DCI requires first the collection of observations of the variables' values at different times. In order to find such sets, in principle, all the possible subsets of system variables should be considered and their DCI computed. In practice, this procedure is feasible only for small-size subsystems in a reasonable amount of time. This paper presents a parallel DCI computation algorithm developed to address this issue.

3   Parallel Algorithm

When large systems are analyzed, the sequential implementation soon reaches unrealistic requirements for computation resources, because the number of possible CRSs increases exponentially with the number of system variables. A possible solution to mitigate this problem consists of a parallel implementation of the main building blocks of the code needed to compute the T_c index.

The GPU is specialized for compute-intensive and highly parallel computation and is capable of addressing highly arithmetically-intense problems that can be expressed as data-parallel computations. The computation of T_c for each CRS is independent of the others, thus a GPU-based parallel code seems to be the most efficient way of computing such an index. That is why we have developed CUDA C code for searching RSs in complex systems.

In order to understand how our code is organized we should consider that the exhaustive computation of the T_c index for all the CRSs of a dynamical system can be divided into the following steps:

1. Computation of the probability distribution function for each system variable;
2. Generation of the homogeneous system;
3. DCI computation for each subset of variables of the homogeneous system;
4. T_c computation for each CRS of the system variables.

From the point of view of the implementation:

- Each sample is stored in a memory area including S adjacent unsigned ints large enough to contain the N_{bit} bits needed to represent the N variables of the system. For example, if we consider a system consisting of N binary variables, then $N_{bit} = N$ and $S = \lceil N_{bit}/\text{sizeof(unsigned int)} \rceil$. If M is the number of samples, then the system data can be stored in an array of $M \cdot S$ unsigned integers.
- Each CRS is represented as a bitmask of N_{bit} bits, where the ith bit is set to 1 if the ith variable is contained in the CRS.

3.1   Computation of the Probability Distribution Function

Each variable of the system is examined individually in order to compute its probability distribution function. In case of binary variables, for example, the distribution of the ith variable is defined by the frequency of the values 0 and 1 (f_{i0} and f_{i1}). The frequency information thus obtained will be used for the generation of the homogeneous system as described in Sect. 3.2.

The frequencies of occurrence of the variables are also used to compute the entropy of each variable, necessary for the computation of the DCI as described in Sect. 3.3. If we consider a binary variable, then the entropy is defined by:

$$H_i = -f_{i0} \cdot log_2 f_{i0} - f_{i1} \cdot log_2 f_{i1}$$

3.2   Homogeneous System Generation

The homogeneous system (HS) is generated from N random variables, homogeneously correlated with one another, having the same probability distribution as the corresponding variables of the dynamical system to be studied.

We obtain M samples by assigning to the ith variable, for each sample, a randomly generated value from the previously estimated distribution.

In case of a system described by binary variables, the ith variable of the homogeneous system, for each sample, will be 0 with probability f_{i0} and 1 with probability f_{i1}. In this way, the HS meets the homogeneity requirement while, at the same time, it maintains a relationship with the dynamical system under consideration.

3.3 DCI Computation on the Homogeneous System

All possible CRS sizes (or classes) from 2 to $N - 1$ are analyzed in order to compute, for each of them, the mean value and the standard deviation of the DCI. If the considered size is r, then the CRSs to be examined are selected by scanning all possible permutations of an N-bit string containing r bits set to 1 and $N - r$ bits set to 0.

The selected CRSs are grouped into grids of T threads each, where each thread is responsible for computing the DCI of one CRS. We have $T = N_B N_T$, where N_B is the number of blocks per grid and N_T is the number of threads per block. Each CRS of a certain size is coupled with its complementary cluster, whose entropy is necessary for computing the mutual information. In other words, each grid is composed of $T/2$ complementary CRS pairs. By synchronizing the execution of parallel threads in order for the entropy of one CRS to be available at the right time, it is possible to compute the statistics of classes r and $N - r$ at the same time, halving the computation time with respect to the original algorithm.

The outputs of this processing step are the mean value and the standard deviation of the DCI for each CRS class of the homogeneous system, which are necessary for computing the T_c index.

3.4 T_c Computation

In the following, we describe the main modules involved in the computation of the T_c index, as shown in Fig. 1.

DCI Module: The computation of the DCI of a CRS consists of three phases:

1. *Creation of the frequency histogram*; the number of occurrences of each value of the CRS is counted; the result is a list of value/number of occurrence pairs;
2. *Entropy computation*; based on the list obtained in the previous phase, the entropy is computed according to Eq. 1;
3. *Computation of the final output*; the threads of the block are synchronized to make the complementary entropy available to each CRS. This enables the computation of the mutual information, which, along with the integration, is used to compute the DCI.

Calculating the frequency histogram is, computationally, the heaviest step. In particular we need:

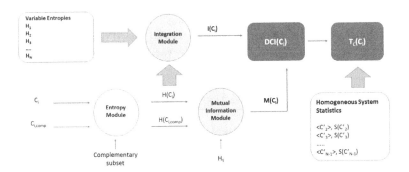

Fig. 1. T_c computation.

– Processing resources to extract the value of the variables in the CRS from each sample of the system;
– Memory to store the frequency histogram of the CRS.

To obtain a good trade-off between performance and memory usage, we generate a hash map, pre-allocated for each thread to be managed by the GPU kernel that computes the histogram (Sect. 3.5).

T_c Module: The module that computes the T_c statistical index is a simple extension of the one which computes the DCI, that takes advantage of the above-mentioned organization into coupled threads. Particularly, in this case, the CRSs of each class, ranging from 2 to $N/2$, are placed aside their complementary CRSs and are inserted, as for the DCI computation for the homogeneous system, in parallel computation batches, each composed of T threads. Once the DCI has been computed, it is sufficient to normalize it according to the statistics (expected value and standard deviation) of the homogeneous system that were obtained earlier (see Sect. 3.3). As the T_c module simply extends the DCI module, both call the same CUDA kernel to perform their computations; the calls differ only in the input parameters.

Once the T_c indices of all the CRSs of the system are obtained, they are compared to select the CRSs having the highest index values.

3.5   Resource Occupation and Scalability

If N is the number of variables that compose the system, the total number of possible CRSs is 2^{N-1}. Thus, the computational complexity of the problem is $O(2^N)$. Parallelizing the computation allows one to obtain a relevant reduction of the execution time. However, this is still not enough to perform an exhaustive search on systems characterized by a large number of variables.

Different considerations can be made regarding memory occupation. Our implementation is based on a simple fact: it is not possible for a CRS to assume a number of configurations that is higher than the number of available samples M, which is usually much lower than the total number of possible CRS configurations (i.e., $M \ll 2^N$). Thus, for each CRS it is possible to pre-allocate a hash table with maximum size M. For this reason, the device memory that is necessary to contain the hash tables of a grid of threads is directly proportional to three independent variables, namely:

- T: number of threads per grid;
- N_{bit}: number of bits needed to store a sample;
- M: number of samples.

Accordingly, the memory occupation increases linearly with the problem size. A good estimation of the device memory needed is:

$$MEM_{TOT} = M \cdot T \cdot (S + 2) \cdot \text{sizeof(unsigned int)} \tag{8}$$

where S is the number of unsigned int that are necessary to store N_{bit} bits. On a device provided with 2 GB of memory, it is possible, for example, to launch 1024 parallel threads and compute the T_c of the same number of CRSs from a system characterized by 1000 binary variables, with $M = 10000$ available samples (in this case, $MEM_{TOT} \simeq 1.4$ GB).

These considerations show that, to analyze large systems, the exponential dependence on the problem size makes an exhaustive search computationally unfeasible. However, an approach based on a metaheuristic would definitely be, as the device memory occupation scales linearly with the problem size.

4   Experimental Results

In this section we illustrate the experimental results we have obtained on four different dynamical systems. The algorithm was evaluated on both artificial and biological systems.

The first case study (referred to as LF) is described by 10 variables and consists of three independent groups, each of which replicates a simple leader-followers dynamic. The model abstracts situations where agents modify their opinion agreeing with (or contrasting the) opinion of other specific agents, called leaders. The system is simply composed of a vector of 10 binary variables $x_1, x_2, ..., x_{10}$ that represent, for example, the positive or negative opinion of 10 agents about a given proposal. The model generates a series of 10 binary vectors (each vector representing an observation of the system) according to the following rules:

- variables are divided into three groups, $G1 = [x_1, ..., x_3]$, $G2 = [x_4, ..., x_6]$ and $G3 = [x_7, ..., x_{10}]$;
- x_1 is a leader; at each step its value is a random value in $\{0, 1\}$;
- the values of the followers x_2 and x_3 are set as a copy of x_1 with probability $1 - p_{noise}$ and randomly with probability p_{noise};
- x_4 and x_7 are "second order" leaders; in each step their values are randomly assigned in $\{0, 1\}$ with probability $1 - p_{copy}$; otherwise x_4 is a copy of x_1 and x_7 is a copy of x_4;

– the values of the followers x_5 and x_6 are set as a copy of x_4 with probability $1 - p_{noise}$ and randomly with probability p_{noise};
– the values of the followers x_8, x_9 and x_{10} are set as a copy of x_7 with probability $1 - p_{noise}$ and randomly with probability p_{noise}.

It is therefore possible to tune the integration among elements in $G1$, $G2$ and $G3$ and the mutual information between $G1$ and $G2$, and between $G2$ and $G3$ by changing p_{noise} and p_{copy} [2,12].

The second and third cases model simplified gene regulatory networks. In particular, the second case study (referred to as AT) models the gene regulatory network shaping the developmental process of Arabidopsis Thaliana; although the whole network is largely unknown, a certain subsystem has been identified as responsible for the floral organ specification. The network is modeled by means of a Boolean network described in [1], having 15 nodes and 10 different attractors (all fixed points): in order to perform an analysis we built a data series containing a number of repetitions of these attractors proportional to the size of their basins of attraction.

The third case (referred to as TH) features 23 Boolean variables, used in [5] to model the regulatory network controlling the T-helper cell differentiation; also in this case we built a data series containing a number of repetitions of the Boolean system attractors proportional to the size of their basins of attraction. We will not discuss about the adequacy of these simplified models, but we will take them for granted and apply our method to test whether it can discover significant MDSs (Mesolevel Dynamical Structures).

The fourth case study is a deterministic simulation of a catalytic chemical system (Catalytic Reaction System - CRS - in the following), characterized by 26 variables, in which there are two distinct reaction pathways: a linear chain and an autocatalytic circle. The reactions happen in an open well-stirred chemostat (CSTR) with a constant influx of feed molecules and a continuous outgoing flux of all the molecular species proportional to their concentration. The dynamics of the system is described adopting a deterministic approach whereby the reaction scheme is translated into a set of ordinary differential equations integrated by means of a Euler method with step-size control. The asymptotic state of this system consists of constant concentrations. In order to apply our analysis, however, one needs to observe the feedbacks in action: thus, we perturbed the concentration of some molecules in order to trigger a response (i.e., a series of changes) in the concentration of (some) other species. The perturbations consisted of temporarily setting to zero the concentration of some species after the system reached its stationary state. To analyze the system response we used a three-level coding where, for each species, the digit '0', '1' and '2' stand respectively for "decreasing concentration", "no change" and "increasing concentration" (Fig. 2) [11,12].

The four cases present different dynamics and representations: in particular, the first test case consists of a binary time series, whereas the second and third cases are the juxtaposition of the binary states of several different attractors, and the fourth case is the encoding of a continuously perturbed situation into a

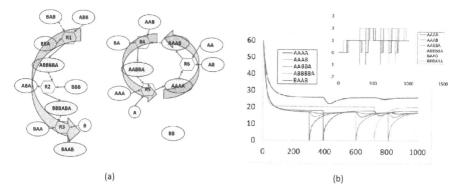

(a)   (b)

Fig. 2. (a) The reaction scheme of the Catalytic Reaction System (CRS): white ellipses represent the chemicals injected in the incoming flux, meshed ellipses represent the chemicals produced inside the CSTR vessel, hexagons represent the reactions; continuous arrows represent the consumptions/productions and dashed arrows represent the catalytic activities. Chemical BB does not participate in any reaction, and it is used as reference. The six reactions are arranged into two independent groups: a linear chain and an autocatalytic circle. (b) A time series of the six produced chemicals and the corresponding three-level encoding.

three-level representation. The method we implemented on GPU is able to find the correct relevant sets in all situations (some of them being discussed in details in [11,12]): in this paper, however, we focus our interest on the performance of the sequential and parallel algorithms.

The parallel algorithm (PA) has been evaluated in terms of correctness and efficiency (speedup), compared to the sequential algorithm (SA). To this purpose, we have used a Linux server provided with CPU Intel(R) Xeon(R) 2.10 GHz, 64 GB of RAM and a GPU NVIDIA GeForce GTX 1070. We have executed 10 independent runs for each example, using different random seeds when generating the homogeneous system.

Table 1 summarizes the algorithms' performance in relation to the system size (expressed as number of variables) and to the number of samples.

In all these case studies the results are correct: they are equal to the ones obtained by the sequential implementation, but they have been computed in a

Table 1. Performance summary of the sequential (SA) and parallel (PA) algorithms

System	#Variables	#Samples	Time (SA)	Time (PA)	Speedup
LF	10	500	2.15 s	11 ms	195.5
AT	15	5000	861 s	0.23 s	3743.5
TH	23	5000	60 h[a]	27.5 s	7854.5
CRS	26	751	20 d[a]	245 s	7053.1

[a] Estimated

significantly shorter time. Using our parallel implementation we can now exhaustively analyze systems of up to 35 variables in less than 24 h.

The speedup with respect to the sequential CPU implementation is very relevant.

5 Conclusion and Final Remarks

In this paper we have presented a fine-grained parallel implementation of the main building blocks needed to compute the T_c index. In summary, the most relevant choices, aiming at algorithm efficiency, are:

– Subgroup-wise parallelization (as opposed to a possible system data-wise parallelization);
– "Smart" allocation of threads/data (like using a hashmap for each thread, implemented on the graphics device).

These choices produce an algorithm which obtains a large speedup, but they are a little more critical as concerns memory allocation.

In the benchmarks we took into consideration, the algorithm obtained a dramatic speedup with respect to the sequential implementation, allowing us to detect RSs in dynamical systems of much larger size than previously possible.

When large systems are analyzed, the increasing number of CRSs makes it impossible to compute the T_c index for every possible subset, even using massively parallel hardware such as GPUs, so we need to design efficient strategies to quickly identify the most promising subsets, limiting the extension of the search.

Considering multi-GPU implementations, the structure of the parallel algorithm is such that the computation of each T_c index is totally independent of the others, which suggests that the number of T_c computations scales almost perfectly linearly with the number of GPUs.

Smart and efficient search algorithms can be easily designed upon our parallel implementation. For example, in [7], we proposed a metaheuristic based on a genetic algorithm that draws the search towards the basins of attraction of the main local maxima in the search space, along with a local search that improves the results by exploring those regions more finely and extensively. Such a metaheuristic computes the fitness function using the GPU-based implementation of the T_c computation described in this paper. The speedups achieved by our parallel implementation of the metaheuristic made it possible for us to analyze systems consisting of up to 137 variables in a reasonable time. Using an exhaustive approach based on a sequential implementation, the same time would have allowed us to analyze only very simple and rather uninteresting systems.

References

1. Chaos, A., Aldana, M., Espinosa-Soto, C., de León, B.G.P., Arroyo, A.G., Alvarez-Buylla, E.R.: From genes to flower patterns and evolution: dynamic models of gene regulatory networks. J. Plant Growth Regul. **25**, 278–289 (2006)

2. Filisetti, A., Villani, M., Roli, A., Fiorucci, M., Poli, I., Serra, R.: On some properties of information theoretical measures for the study of complex systems in advances in artificial life and evolutionary computation. Commun. Comput. Inf. Sci. **445**, 140–150 (2014)
3. Filisetti, A., Villani, M., Roli, A., Fiorucci, M., Serra, R.: Exploring the organisation of complex systems through the dynamical interactions among their relevant subsets. In: Andrews, P., et al. (ed.) Proceedings of the European Conference on Artificial Life 2015 (ECAL 2015), pp. 286–293. The MIT Press (2015)
4. Gershenson, C., Fernandez, N.: Complexity and measuring emergence, self-organization, and homeostasis at multiple scales. Complexity **18**(2), 29–44 (2012)
5. Mendoza, L., Xenarios, I.: A method for the generation of standardized qualitative dynamical systems of regulatory networks. Theor. Biol. Med. Model. **3**(1), 13 (2006)
6. Prokopenko, M., Boschetti, F., Ryan, A.J.: An information-theoretic primer on complexity, self-organization, and emergence. Complexity **15**(1), 11–28 (2009)
7. Sani, L., et al.: Efficient search of relevant structures in complex systems. In: Adorni, G., Cagnoni, S., Gori, M., Maratea, M. (eds.) AI*IA 2016. LNCS (LNAI), vol. 10037, pp. 35–48. Springer, Cham (2016). doi:10.1007/978-3-319-49130-1_4
8. Shalizi, C.R., Camperi, M.F., Klinkner, K.L.: Discovering functional communities in dynamical networks. In: Airoldi, E., Blei, D.M., Fienberg, S.E., Goldenberg, A., Xing, E.P., Zheng, A.X. (eds.) ICML 2006. LNCS, vol. 4503, pp. 140–157. Springer, Heidelberg (2007). doi:10.1007/978-3-540-73133-7_11
9. Sporns, O., Tononi, G., Edelman, G.: Theoretical neuroanatomy: relating anatomical and functional connectivity in graphs and cortical connection matrices. Cereb. Cortex **10**(2), 127–141 (2000)
10. Tononi, G., McIntosh, A., Russel, D., Edelman, G.: Functional clustering: identifying strongly interactive brain regions in neuroimaging data. Neuroimage **7**, 133–149 (1998)
11. Villani, M., Filisetti, A., Benedettini, S., Roli, A., Lane, D., Serra, R.: The detection of intermediate level emergent structures and patterns. In: Liò, P., Miglino, O., Nicosia, G., Nolfi, S., Pavone, M. (eds.) Proceedings of ECAL2013, The 12th European Conference on Artificial Life. MIT Press (2013)
12. Villani, M., Roli, A., Filisetti, A., Fiorucci, M., Poli, I., Serra, R.: The search for candidate relevant subsets of variables in complex systems. Artif. Life **21**(4), 412–431 (2015)

Complexity Science for Sustainable Smart Water Grids

Angelo Facchini[1,2]([✉]), Antonio Scala[1,2], Nicola Lattanzi[3], Guido Caldarelli[1,2], Giovanni Liberatore[4], Lorenzo Dal Maso[5], and Armando Di Nardo[6]

[1] IMT-Alti Studi Lucca, Lucca, Italy
4.facchini@gmail.com
[2] CNR-Istituto dei Sistemi Complessi, Roma, Italy
[3] Università di Pisa, Pisa, Italy
[4] Università di Firenze, Florence, Italy
[5] Erasmus University Rotterdam, Rotterdam, The Netherlands
[6] Seconda Università degli Studi di Napoli, Caserta, Italy

Abstract. While the effects of climate change unfold and become more visible, infrastructures – especially those related to the distribution of water and energy – are the most exposed to the deep changes expected in the next years. Water is fundamental for people, and for infrastructures like energy, waste, and food production. Water sustainability is therefore a fundamental aspect to be addressed by an efficient use of the resources and by mainteining high quality standards. Hence, water industry and water infrastructure need a deep transformation; in this paper we present a framework based on complex systems and management science as a possible pathway to reshape and optimize the performance of the water infrastructure to cope with the complexity of todays' challenges. To this aim, we propose the framework *Acque 2.0* (Water 2.0), where we point out how the increase of the infrastructural resilience and of the overall quality of service can be attained by integrating models, algorithms and numerical methods like network simulations and big data analytics for the predictive maintenance of water networks. We discuss how Complexity Science is the natural glue allowing technical, management and social issues to be integrated in the holistic vision of the "water system" needed play to provide measures for an integrated sustainability reporting that involves utilities, regulators, policy makers, and citizens.

1   Introduction

Resources (water, energy, materials) are scarce and the environment can not be longer considered an infinite reservoir for our needs and a infinite sink for our waste, whether they are wastewater or air pollutants such as fine particulates or greenhouse gases. Pressure on the environment is caused by several factors like the increase of population or the fast development of Asian and African countries; in particular, urban development is possibly one of the most relevant. In fact, according to the UN [1] and McKinsey Global Institute [2], since urban population is expected to increase, there is the urgent need to cope with a

© Springer International Publishing AG 2017
F. Rossi et al. (Eds.): WIVACE 2016, CCIS 708, pp. 26–41, 2017.
DOI: 10.1007/978-3-319-57711-1_3

multi-faceted challenge that involves environment, efficient resource use, social and economic systems and to develop a more efficient and sustainable way to guarantee the current standard for basic infrastructural services (e.g. distribution of water and energy, mobility, and waste treatment/collection). In this paper we will focus on water.

Despite the fact that fresh water is the most essential resource for mankind, it is scarce and continuously declining around the world due to factors like climate change, competition for uses, population growth, urbanization, agricultural activities and industrialization [3]. For these reasons, water sustainability has become one of the prominent topics in the agenda of policy and decision makers [4] and has been quantified using performance indicators for reliability, resiliency, and vulnerability. As an example, Tabesh [5] proposed a model using hydraulic, physical, and empirical indices for the rehabilitation of water distribution networks, while Pirlata [6] modeled a sustainable water distribution system considering trade-offs between hydraulic reliability, life cycle cost and CO_2 emissions; Li [7] defined sustainability in water systems as an equilibrium between network efficiency and resiliency. In the last years – especially in the fields of telecommunication and energy – researchers and utilities are converging towards an integrated management of infrastructures [8], interweaving elements of asset management, sustainability reporting and the modern methods of data analysis and collection (e.g. big data).

In this paper we present an integrated framework for sustainable water infrastructures based on complex systems, management science and sustainability reporting. In particular, we think that the adoption of algorithms and numerical methods based on a holistic view of the water infrastructure will allow for a full development of water infrastructures steering a sustainability transition.

The paper is organized as follows: In Sect. 2 we describe a multidisciplinary approach discussing the advantages of using the complex systems framework and a description of water infrastructures based on complex networks. A set of case studies and best practices developed by researchers is used to describe how resilience, vulnerability analysis and multi-criteria optimization can be exploited to increase the infrastructure's sustainability. Section 3 is devoted to integrated sustainability reporting also referring to the case of the Italian legislation. Finally we state our conclusions in Sect. 4.

2   A Multidisciplinary Approach to Sustainable Water Infrastructures Management

We propose a holistic approach called *Acque 2.0* (Water 2.0) that aims to fully exploit the sustainability potential of water infrastructures by means of the following elements:

1. Smart infrastructures, allowing for a real-time description of the status of the networks and loads;

2. Complex Systems, including algorithms and visions allowing for the improvement of quality of service and resilience.
3. Sustainability reporting and performance measures.

2.1 Smart Water Infrastructures

In the field of electric power distribution systems, smart grids represented a revolution, both at the micro-scale (installation of electronic meters) and at the macro-scale (automation of medium and high voltage substations). The benefits of digitalization in distribution networks are well evidenced by the success achieved by the electric utility Enel, that in last two decades installed in Italy about 4 million electronic meters that have enabled a marked improvement in network management and operational efficiency, and ensured high quality standards to customers while maintaining sustainable operating costs [9].

In the last years, both regulators and utilities are working for the advent of smart water metering [10]. In fact, water utilities – especially those operating in water scarce conditions – need to reduce the impact on resources while reducing operating costs and ensuring high quality standards of service. Furthermore, smart metering offers the following advantages:

1. Reduction of technical and administrative losses
2. Improvement of prediction capabilities for the peak demand
3. Reduced costs associated with meter reading and operating costs
4. Improved response times in case of failures
5. Accurate monitoring of quality of service (e.g., monitoring of pollutants)
6. Improved quality of services and possibility of developing new business models

The advances in metering and data communications technology have made it possible to record household water usage data through smart water meters. Such devices can automatically and electronically capture, collect and communicate water usage by real time (or close to real time) readings. These electronic data can be transferred by automated means (e.g. GSM, GPRS, CDMA, drive by) to servers for storage and for the subsequent processing and analysis of data [11]. Smart water metering would be expected at least to convey daily meter readings between the water utility and the water meter, and potentially to customers as well. Finer levels of data capture (in seconds, minutes or hourly) could also be programmed into the loggers to enable more detailed analysis to be carried out (e.g. [12]). Such an approach is unlike traditional methods of periodical (accumulation) metering, where household water consumption is typically "manually" read on a monthly or quarterly basis, forcing the daily trends of consumption needed for planning purposes to be inferred indirectly. Instead, automated metering would provide benefits both for water authorities and consumers in monitoring and controlling water consumption [13], possibly enabling alternative pricing mechanisms such as time-of-use or seasonal tariffs [14,15]. Real time operations and monitoring also allow for the use of modern analysis methods based on complex systems; in particular, in the following section we describe a complex systems' approach to water infrastructures.

2.2 Complex Systems, Nonlinear Time Series, and Complex Networks

Complex systems and nonlinear methods are, by now, well established and widely described in a rich literature (see [16] and citations therein). The fact that apparently simple deterministic systems may exhibit complicated temporal behaviors in the presence of nonlinearity has influenced thinking and intuition in many fields. In particular, nonlinear methods have been successfully applied to a wide range of natural phenomena, giving insights and providing solution in different disciplines. Within nonlinear methods, nonlinear analysis of time series plays a fundamental role when analyzing experimental data, especially when mathematical models are hard to develop or give only poor information to the experimentalist [17]. The main task of nonlinear time series analysis (NTSA) is to extract information on the nonlinear system, assuming the hypothesis that a single or a multivariate recording represents the evolution of a unknown dynamical system (i.e. a systems described by a set of nonlinear differential equations) and its past evolution contains information about the (unknown) model that has produced the time series itself [18].

In the last two decades researchers successfully characterized a wide range of phenomena like mechanical systems, markets (including energy and commodities), biological and biophysical systems, ecology etc. (the reader is referred to Abarbanel [19] and Kantz [17] for an extensive description of methods and their applications).

With regards to their structure, complex systems have been successfully characterized by means of networks. Indeed, in physics, a network is any real system that can be represented by mathematical objects called *graphs*. A graph is defined by a set of *vertices* (also called *nodes*) and a set of connections, between them, called *edges* (or *links*). Edges connecting vertices can be alternatively *undirected*, if a preferential direction is not defined on them, or *directed*, if a preferential orientation is present. A graph built by directed edges is called a directed graph. It is also possible associate a certain value to an edge to take into account the load carried by that edge; in this case we are in front of a *weighted graph*. A graph composed by n vertices and m edges is usually indicated by $G(n, m)$. The two quantities n and m are called *order* and *size* of the graph respectively, and they are not independent of each other: for undirected graphs the maximum value of the size is $m = n(n-1)/2$ while for directed graphs yields $m = n(n-1)$. The structure of a graph $G(n, m)$ can be also represented by an adjacency matrix $A(n, n)$ whose entries a_{ij} are 1 if there is an edge connecting i to j and 0 otherwise. In a weighted graph, the entries different from 0 consist in real numbers accounting for the weight associated to the edge. For undirected graphs, the adjacency matrix is symmetrical. Figure 1 shows a directed graph of *order* 5 and *size* 5, directions are indicated by the arrows. The adjacency matrix is:

$$a_{ij} = \begin{pmatrix} 0 & 0 & 0 & 1 & 1 \\ 1 & 0 & 0 & 1 & 0 \\ 0 & 0 & 0 & 0 & 0 \\ 0 & 0 & 1 & 0 & 0 \\ 0 & 0 & 0 & 0 & 0 \end{pmatrix} \tag{1}$$

where a semicolon indicates the end of a matrix row. The undirected graph corresponding to the topology of Fig. 1 leads to the symmetrical adjacency matrix

$$\begin{pmatrix} 0\,1\,0\,1\,1 \\ 1\,0\,0\,1\,0 \\ 0\,0\,0\,1\,0 \\ 1\,1\,1\,0\,0 \\ 1\,0\,0\,0\,0 \end{pmatrix} \tag{2}$$

If in a graph of order n all possible edges are present, the graph is called *complete* and is indicated by K_n while, if no edge is drawn the graph is called *empty* and is indicated by E_n.

Fig. 1. Example of a directed graph $G(5,5)$ of order 5 and size 5. The arrows show the edge directions.

Starting from the adjacency matrix a set of measures can be defined, for an exhaustive survey on complex networks and their applications the reader can refer to the review article by R. Albert and A.L. Barabasi [20], and the book by Caldarelli [21]. In the following the most representative are briefly described:

- the *degree* k_i of a vertex i is the number of edges attached to it; in an undirected graph any edge contributes to the degree of each of the two vertices it connects; it is not the same for directed graphs, where one can define an *in-degree* and an *out-degree* with an obvious meaning; the list of the degrees of all vertices present in a graph is called a *degree sequence*.
- in the case of *weighted graphs*, one can define also a generalization of the concept of *degree*, which is the strength or *weighted* degree of a vertex, k_i^w: it is calculated as the sum of the weights carried by the edges attached to the node. The definitions of *in-strength*, *out-strength* and *strength* sequence are straightforward;
- in an undirected graph two vertices are *connected* if there is a *path* (i.e. a sequence of edges) between them. If for every couple of vertices there is a path, the graph is called *connected*. A *connected component* (CC) is a connected sub-graph in a graph. In a directed graph one can have *strongly* and *weakly* connected components (SCC and WCC respectively). In the former case directed paths exist for every pair of vertices, while in the latter case paths exist only when considering the edges as undirected.

Increase of Infrastructure Resilience Using Complex Networks. Recent academic studies have shown how very different infrastructural networks (e.g., water, electricity, gas, transport, etc.) or "soft" networks of relationships (e.g., bank loans or trade flows) share a general topological structure (topology) [22]. In particular, complex networks are able to describe such heterogeneous systems as a graph that can be described by the above mentioned measures, allowing for:

- Network optimization with respect to multiple cost functions
- Network stability to failures or sabotage
- Measure the impact of failures on the population

By using the complex networks methods, the analysis of water infrastructures can be organized in the following steps:

1. Characterization in the form of weighted graph of the water distribution network;
2. Definition of cost functions (e.g. minimizing use sources, electricity consumption to serve the area, quality of service);
3. Identification of critical nodes obtained by combining network topology data (e.g., the node centrality) with historical data on the working life of the network components;
4. Definition of a complex network model that takes into account the historical infrastructure of the actual distribution network.
5. Overall risk assessment by taking into account interactions between individual risks based on the complex network model.
6. Implementation of a *predictive maintenance algorithm* based on historical data and network measures.

Topological and Vulnerability Analysis. Complex Systems are often organized in the form of networks whose nodes and arcs are located in ordinary space. The current state of the structure and evolution of such spatial networks is summarized in [23]. The first method to describe and understand a network is the so-called topological analysis that analyzes the properties of a network starting only by the adjacency relationships (i.e. who is connected to whom). The starting point of topological analysis is the observation that many networks show a surprising degree of fault tolerance, often at the expense of extreme vulnerability to attacks targeted at nodes/arcs of particular importance [24]. A possible first application is the use of a set of advanced analysis techniques to study the topological connectivity of the water distribution system to understand the robustness and susceptibility to damage. In this phase, the water distribution systems are modeled as weighted and direct networks using the physical attributes and hydraulic system components. One can then use descriptive metrics to quantify the properties of a water system both locally (individual components) and globally (water flows); in addition, these metrics allow to support the identification of critical nodes and their rankings compared to impacts. The relevance of this approach has already been demonstrated through a series of case studies on local water systems in the US and UK [25] (see Fig. 2).

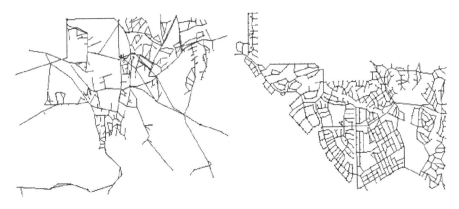

Richmond's Top Ten Critical Nodes

Node	Node Degree	Base Demand (L s^{-1})	Demand-Adjusted Entropic Degree	Normalized Demand-Adjusted Entropic Degree
N20	2	0.33	94.37	1.00
N219	2	0.36	75.59	0.80
N208	2	0.32	64.38	0.68
N153	2	0.02	54.51	0.58
N97	3	0.30	48.35	0.51
N120	2	0.28	41.38	0.44
N131	3	0.01	36.13	0.38
N9	2	0.18	31.15	0.33
N31	2	0.15	26.28	0.28
N230	2	0.19	24.48	0.26

Colorado Springs' Top Ten Critical Nodes

Node	Node Degree	Base Demand (L s^{-1})	Demand-Adjusted Entropic Degree	Normalized Demand-Adjusted Entropic Degree
N144	4	0.53	542.28	1.00
N1229	3	0.53	354.36	0.65
N1373	3	0	192.27	0.36
N126	3	0.25	146.34	0.27
N3	4	0.25	125.04	0.23
N215	4	0.53	122.07	0.23
N143	4	0.53	111.27	0.21
N1540	3	0.53	109.30	0.20
N1542	3	0.53	97.70	0.18
N315	3	0.53	90.88	0.17

Fig. 2. Upper panel: Example of weighted graph representation of a water distribution network (see [25] for additional details). Lower panels: Top ten critical nodes for the case studies considered in [25].

Vulnerability can be also assessed by means of *percolation* analysis, an approach addressing the purely geometrical problem of the connectivity of a set of spatially distributed points according to a pre-defined distribution [26]. Notice that in computer systems and networks, the quantification of the probability that one node can reach (i.e. is connected) to another is called reliability [27]. Hence, percolation models are the first in the row to be used in a robustness analysis since they work on a purely topological level provided giving information on the robustness of the system respect to the service interruption probability for single pipes. A possible application of the percolative approach for reliability assessment of urban and suburban distribution networks is based on the component analysis, on the network topology, and on the survival analysis of the elements; in fact, in water networks the effects of past performance of the

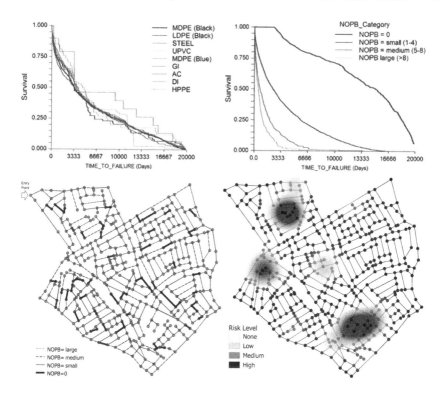

Fig. 3. Upper panels: (left) Survival curves for pipes according to the construction material; (right) Survival curves for concrete pipes according to NPOB. Lower panels: (left) Topology of the distribution network (right) spatial analysis (heat-map) of network reliability. Darker zones show higher failure risk [28].

network by including the Number of Previous Observed breaks (NOPB) are a risk factor that enhance the probability of failure of a pipe. Such an approach has been successfully applied to investigate the effects of the past performance of a network on the seismic and/or not-seismic reliability evaluation [28] (Fig. 3).

Analysis of Systemic Failures (cascadings) in Water Networks. Following the introduction of the free market, in recent years infrastructures have been subject to relevant stress due to the increased demand and the need to reduce and/or plan the operational expenses. Even in the case of the extremely robust electric networks, this phenomenon has led to the increase of the cascade phenomena (blackouts). In particular, the load condition of the lines is an important indicator of possible imminent breakage [29]. In the case of water supply, one can obtain the load profile using network data modelled by means of EPANET (http://epanet.de/), a freely available software for modeling water networks [30]. Consequently, one could apply the same methodology used to analyze the falls of electrical networks in the case of water infrastructure, as done by [31] (Fig. 4).

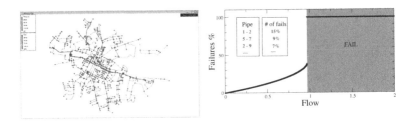

Fig. 4. Hypothesis of cascading failures analysis for water networks: (left) Standard representation of pressure map (right) Weak point analysis.

Water Networks Partitioning. Water budget to reduce water losses is achieved with Water Network Partitioning that consists in dividing a water network in k smaller subsystems (or District Meter Areas, DMAs) by gate valves and flow meters [32]. WNP allows improving the management and protection of water distribution systems [33] although an innovative tool to define optimal DMA is required since the insertion of gate valves in the water network can reduce significantly the hydraulic performance in terms of water pressure [33]. Indeed, the closure of some pipes reduces the hydraulic section and, consequently, increases the head losses. Traditionally, water network partitioning was achieved through empirical approaches based on trial and error procedures and literature suggestions [32]. Anyway, these approaches do not allow to find the optimal solution because the computational problem is a NP problem. Recently, graph and complex network theory have been successfully used to water network partitioning The procedures are based on different approaches: spectral clustering [34,35]; graph theory [36–38], multi-agent [39]; community detection [40]. An example of Water Network Partitioning of Parete network, a little city located in a densely populated area southern of Caserta (Italy) with 10,800 inhabitants, is reported in the Fig. 5.

Improve Quality of Service by Means of Time Series Methods. According to the experience of electricity distribution networks [31], water demand forecasting is a key point to be addressed to improve the quality of service, especially in regions with strong population variability. Following [31], a possible application of such methods goes in the direction of realizing a short-term water demand forecasting tool for water networks, with the aim to implement a predictive control infrastructure. In particular, a first approach may apply the Gaussian processes for the prediction of very short-term fluctuations joined to a double-Seasonal HoltWinters method (DSHW) for the systematic part. A recent study [41] shows how these methods can be successfully applied to a smart network characterized by a high-level modeling (by example Barcelona water network, as shown in Fig. 6).

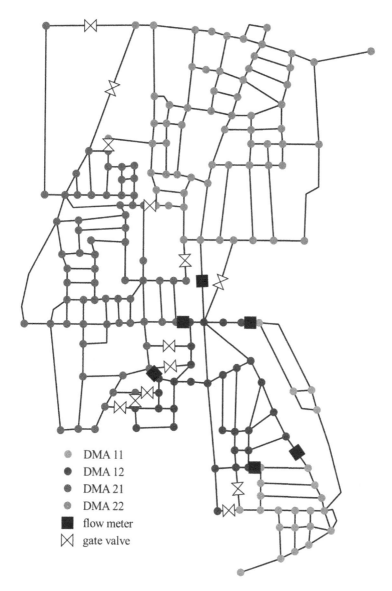

Fig. 5. Parete network without DMA and Water Network Partitioning of Parete network with 4 DMA.

3   Integrated Sustainability Reporting for Water Utilities

In this section we briefly introduce the Integrated Reporting as a part of the interdisciplinary framework Acque 2.0. In our vision, the technical and algorithmic part described in Sect. 2 is not able to describe alone the complexity of water

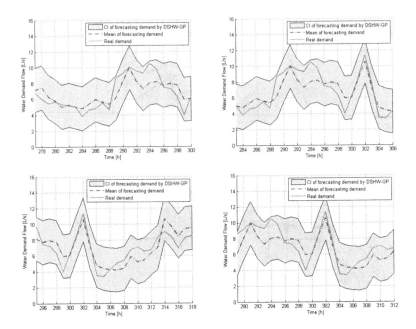

Fig. 6. "Gaussian-Process-based Demand Forecasting for Predictive Control of Drinking Water Networks" [41]: results obtained for the municipality of Barcelona suggest the reliability of the forecasting method.

infrastructures management, that also involves relationships with stakeholders, regulation authorities, clients, and markets.

Sustainability assessments embeds the principles of sustainable development, a notion promoted by the Brundtland Commission in the well known report *Our Common Future* in 1987 [42], and further extended during the 1992 Rio summit conducted by the United Nations Conference on Environment and Development and again at the World Summits on Sustainable Development in 2002 and 2005. Sustainable development is described as "the development that meets the needs of the present without compromising the ability of future generations to meet their own needs", and is based on two key concepts:

1. The concept of "needs", in particular, the essential needs of the world's poor, to which overriding priority should be given;
2. The idea of limitations imposed by the state of technology and social organization on the environment's ability to meet present and future needs.

In the last three decades sustainability science greatly evolved as new knowledge and a new point of view on ecosystems and the socio-economic system [42–46].

Accordingly, companies started to use corporate social responsibility (i.e., CSR) as a vehicle to appear socially aware according to the expectations of stakeholders [47,48] establishing, thus, congruence between corporate operations and social values. Besides, there has been an increasing number of companies that started to voluntary disclose non-financial information related to their commitment to environmental preservation, human rights protection, as well as employees and social welfare [49]. In Europe, the directive on disclosure of non-financial and diversity information (2014/95/EU), entered into force in December 2014, requires large European Union (EU) companies (of over 500 employees) to include in their management reports a non-financial statement on the impact of their activities relating to environmental, social, and employee concerns, including respect for human rights and efforts to combat corruption and bribery.

The concept of sustainability is extended to urban water management as well as dictated in Agenda 21, in the Sustainable Development Goals, and in the recent urban agenda released by UN-Habitat the last October, 2016. A first definition of sustainable urban water can be found in [50]: "a sustainable urban water system should over a long time perspective provide required services while protecting human health and the environment, with a minimum of scarce resources". To clarify our understanding of sustainable urban water services, we begin by defining the urban water system as one that includes collection, treatment and distribution of water, wastewater and stormwater [51].

Considering this point of view, integrated reporting is a suitable mean to assess the sustainability of a resource use and the connected infrastructures. In fact, the primary purpose of an integrated report is to explain to providers of financial capital how an organization creates value over time, benefitting all stakeholders interested in an organization's ability to create value over time, including employees, customers, suppliers, business partners, local communities, legislators, regulators and policy-makers. Furthermore, an integrated report aims to provide insight about the resources and relationships used and affected by an organization, also explaining how the organization interacts with the external environment and the capitals to create value over the short, medium and long term. Furthermore, Integrated Reporting allows firms to "brings together material information about an organizations strategy, governance, performance and prospects in a way that reflects the commercial, social and environmental context within which it operates. It provides a clear and concise representation of how an organization demonstrates stewardship and how it creates and sustains value" [52]. That said, the underling idea is to let firms adopt principles in order to provide information regarding: (a) significant external factors that affect an organization, (b) the resources and relationships used and affected by the organization, and (c) how the organizations business model interacts with external factors and resources and relationships to create and sustain value over time. In doing so, Integrated Reporting provides a meaningful presentation of firms long term resilience and success and facilitate the outflow of information to investors and to other relevant stakeholders.

Regarding water, the protection of resources requires unified and integrated action, rather than action based on isolated sectors, taking into account all the phenomena and all human activities that directly or indirectly affect the quality of water resources and of the related services. According to the OECD "water is the perfect example of a sustainable development challenge encompassing environmental, economic and social dimensions (2003:19). For this reason the EU Water Framework Directive (2000/60/EC)[1] of the European Parliament settled-up a comprehensive framework on water policy, indicating the main elements for the attainment of the ecological quality of water, expanding the view to groundwater and to other related quantitative aspects. In Italy, despite the advancement of WFD, water governance remains overly complex and oriented towards short term problem solving. [...] Italy needs to urgently formulate a strategic vision for the water sector, outlining how the national government can most effectively support regional and local authorities in managing water resources, taking account of territorial disparities in resource endowments, policy priorities and capacities. [...] These efforts should be supported by provision for stakeholder and public participation in decision making, and for transparency and accountability" [53].

It is therefore clear that to boost the intervention potential in the field of water is necessary to develop integrated resource management capabilities by focusing on a performance measurement system that must be able to pursue the following cognitive and organizational goals in line with the European Water Stewardship (EWS) standard (http://www.ewp.eu/):

- Assist and guide in predictions on evolutionary scenarios of water requirements;
- Identify the specific employment analysis parameters of water at the local level;
- Steer the decision-making process towards the selection of targets and building address plans that maximize the degree of compliance with the regulatory framework and authorization locally and nationally;
- Communicate to stakeholders the strategic behavior and attitude to the creation of enterprise value (Integrated Reporting) according to standard IIRC (http://integratedreporting.org/).

4   Conclusions

In this paper we discussed the framework "Acque 2.0" as a set of tools, algorithms and methods – some already present, some to be developed – that in our opinion is suitable to fully exploit the sustainability potential of water infrastructures. To such an aim, an holistic approach is needed; hence, we have based our vision on complex systems and complex networks, whose methods are both able to improve the infrastructure decreasing operational costs (while ensuring a high quality of service) and to provide a set of reliable measures for the integrated sustainability reporting. Because of their structures, the technological advancement of water network is still at its beginning, and methods based on complex systems science, such as complex networks, genetic algorithms, and artificial intelligence

---

[1] Directive2000/60/EC of the European Parliament and of the Council of 23 October 2000 establishing a framework for Community action in the field of water policy at: http://eur-lex.europa.eu/legal-content/EN/TXT/?uri=CELEX:32000L0060.

can be exploited to define new measures and Key Performance Indicators for the optimal management of water networks. Furthermore, new algorithms can be exploited to implement decision systems supporting both utilities and policy makers. By example, complex networks and genetic algorithms may be used to implement network partition strategies to reduce water losses and provide a definition of water districts going beyond the geographical aspects only. Integrated reporting also benefits from a set of complexity-based KPIs because they are able to describe the network as a whole, considering factors emerging from the interaction of multiple components (e.g. water-energy nexus).

Indeed, embracing the Complexity approach means embracing a vision that is systemic and long-term. A vision that is able to transform the infrastructure making it modern, efficient, and at the same time able to meet the needs of users while decreasing its environmental footprint.

We strongly believe that such a process of transformation should be the result of a synergy between all actors involved: utilities, regulators, the research community, and citizens. The utilities, in particular, as the main actors of this transformation, are called to lead it: even powerful frameworks as Complex Systems are empty shells if the targets are not well defined.

References

1. United Nations. World urbanization prospect, the 2014 revision. Technical report (2014)
2. McKinsey Global Institute. Resource revolution. Technical report (2012)
3. UNESCO. Water for a sustainable world. Technical report (2015)
4. Aydin, N.Y., Mays, L., Schmitt, T.: Sustainability assessment of urban water distribution systems. Water Resour. Manage. **28**, 4373–4384 (2014)
5. Tabesh, M., Saber, H.: A prioritization model for rehabilitation of water distribution networks using gis. Water Resour. Manage. **26**, 225–241 (2012)
6. Pirlata, K.R., Ariaratnam, S.T.: Reliability based optimal design of water distribution networks considering life cycle components. Urban Water J. **9**(5), 305–316 (2012)
7. Li, Y., Yang, Z.F.: Quantifying the sustainability of water use systems: calculating the balance between network efficiency and resilience. Ecol. Model. **222**, 1771–1780 (2011)
8. Hansman, R.J., Magee, C., de Neufville, R., Robins, R., Roos, D.: Research agenda for an integrated approach to infrastructure planning, design and management. Int. J. Crit. Infrastruct. **2**(2/3), 146–159 (2006)
9. Onyeji, I., Colta, A., Papaioannou, I., Mengolini, A.: Smart Grid projects in Europe: lessons learned and current developments. JRC (2011)
10. Gurung, T.R., Stewart, R.A., Sharma, A.K., Beal, C.D.: Smart meters for enhanced water supply network modelling and infrastructure planning. Resources **90**, 34–50 (2014)
11. Boyle, T., Giurco, D., Mukheibir, P., Liu, A., Moy, C., White, S., Stewart, R.: Intelligent metering for urban water: a review. Water **5**(3), 1052–1081 (2013)
12. Britton, T.C., Stewart, R.A., O'Halloran, K.R.: Smart metering: enabler for rapid and effective post meter leakage identification and water loss management. J. Cleaner prod. **54**, 166–176 (2013)

13. Stewart, R.A., Willis, R., Giurco, D., Panuwtwanich, K., Capati, G.: Web-based knowledge management system: linking smart metering to the future of urban water planning. Aust. Planner **47**(2), 66–74 (2010)
14. Cole, G., O'Halloran, K.R., Stewart, R.A.: Time of use tariffs: implications for water efficiency. Water Sci. Tech. Water Supply **12**, 90–100 (2012)
15. Parker, J.M., Wilby, R.L.: Quantifying household water demand: a review of theory and practice in the UK. Water Res. man. **27**, 981 (2013)
16. Strogatz, S.: Nonlinear Dynamics and Chaos, 2nd edn. Westview Press, Boulder (2014)
17. Bradley, E., Kantz, H.: Nonlinear time-series analysis revisited. Chaos **25**(9), 097610 (2015)
18. Kantz, H., Schreiber, T., Hegger, R.: Nonlinear Time Series Analysis. Cambridge University Press, Cambridge (2005)
19. Abarbanel, H.D.I.: Analysis of Observed Chaotic Data. Springer, New York (1997)
20. Albert, R., Barabsi, A.L.: Statistical mechanics of complex networks. Rev. Mod. Phys. **74**, 47–97 (2002)
21. Caldarelli, G.: Scale Free Networks: Complex Webs in Nature and Technology. Cambridge University Press, Cambridge (2007)
22. D'Agostino, G., Scala, A.: Networks of Networks: The Last Frontier of Complexity. Springer, Switzerland (2014)
23. Reka, A., Jeong, H., Barabasi, A.L.: Error and attack tolerance of complex networks. Nature **406**, 378–382 (2000)
24. Barthelemy, M.: Spatial networks. Phys. Rep. **499**, 1–101 (2011)
25. Yazdani, A., Jeffrey, P.: Applying network theory to quantify the redundancy and structural robustness of water distribution systems. J. Water Resour. Plann. Manage. **138**(2), 153–161 (2012)
26. Ángeles Serrano, A., Boguñá, M.: Clustering in complex networks. ii. percolation properties. Phys. Rev. E **74**, 056115 (2006)
27. Shooman, M.L.: Reliability of Computer Systems and Networks: Fault Tolerance, Analysis, and Design. Wiley, New York (2002)
28. Christodoulou, S., Fragiadakis, M.: Vulnerability assessment of water distribution networks considering performance data. J. Infrastruct. Syst., 04014040 (2015)
29. Sakshi Pahwa, Caterina Scoglio, and Antonio Scala. Abruptness of cascade failures in power grids. Sci. Rep., 4:-, January 2014
30. Rossman, L.A.: EPANET2 Users Manual. US E.P.A., Cincinnati (OH) (2000)
31. Mureddu, M., Caldarelli, G., Chessa, A., Scala, A., Damiano, A.: Green power grids: How energy from renewable sources affects networks and markets. PLoS ONE **10**(9), e0135312 (2015)
32. Water Authorities Association and Water Research Centre. Leakage control policy and practice. technical working group on waste of water. Technical report (1985)
33. Di Nardo, A., Di Natale, M., Santonastaso, G.F., Venticinque, S.: An automated tool for smart water network partitioning. Water Resour. Manage. **27**(13), 4493–4508 (2013)
34. Herrera, M., Canu, S., Karatzoglou, A., Péres-garcia, R., Izquierdo, J.: An approach to water supply clusters by semi-supervised learning. In: Swayne, D.A., Yang, W., Voinov, A.A., Rizzoli, A., Filatova, T. (eds.) Proceedings of International Environmental Modelling and Software Society (IEMSS) (2010)
35. Di Nardo, A., Di Natale, M., Giudicianni, C., Greco, R., Santonastaso, G.F.: Water supply network partitioning based on weighted spectral clustering. In: Cherifi, H., et al. (eds.) Complex Networks & Their Applications V. SCI, vol. 693, pp. 797–807. Springer, Cham (2016)

36. Sharma, A.K., Swamee, P.K.: Application of flow path algorithm in flow pattern mapping and loop data generation for a water distribution system. J. Water Supply Res. T. **54**(7), 411–422 (2005)
37. Gomes, R., Sá Marques, A., Sousa, J.: Identification of the optimal entry points at district metered areas and implementation of pressure management. Urban Water J. **9**(6), 365–384 (2012)
38. Di Nardo, A., Di Natale, M., Santonastaso, G.F.: A comparison between different techniques for water network sectorization. Water Sci. Technol. Water Supply **14**(6), 961–970 (2014)
39. Izquierdo, J., Herrera, M., Montalvo, I., Pérez-Garca, R.: Division of water distribution systems into district metered areas using a multi-agent based approach. Comm. Com. Inf. S.C. 50(4), 167–180 (2011)
40. Diao, K., Zhou, Y., Rauch, W.: Automated creation of district metered area boundaries in water distribution systems. J. Water Res. P.L.-ASCE **139**(2), 184–190 (2013)
41. Wang, Y., Ocampo-Martínez, C., Puig, V., Quevedo, J.: Gaussian-process-based demand forecasting for predictive control of drinking water networks. In: Panayiotou, C.G.G., Ellinas, G., Kyriakides, E., Polycarpou, M.M.M. (eds.) CRITIS 2014. LNCS, vol. 8985, pp. 69–80. Springer, Cham (2016). doi:10.1007/978-3-319-31664-2_8
42. United Nations World Commission on Environment and Development. Our Common Future. Oxford University Press (1987)
43. Costanza, R., Daly, H., Bartholomew, J.A.: Goals, agenda, and policy recommendations for ecological economics. In: Costanza, R. (ed.) Ecological Economics, pp. 1–21. Columbia University (1991)
44. Pearce, D.W., Atkinson, G.D.: Capital theory and the measurement of sustainable development: an indicator of sustainability. Ecol. Econ. **8**, 103–108 (1993)
45. Costanza, R., Patten, B.C.: Defining and predic- ting sustainability. Ecol. Econ. **15**, 193–196 (1995)
46. Schubert, A., Láng, I.: The literature aftermath of the brundtland reportour common future a scientometric study based on citations in science and social science journals. Environ. Dev. Sustain. **7**, 1–8 (2005)
47. Ioannou, I., Serafeim, G.: What drives corporate social performance? the role of national-level institutions. J. Int. Bus. Stud. **43**, 834–864 (2012)
48. Cho, C.H., Laine, M., Roberts, R.W., Rodrigue, M.: Organized hypocrisy, organizational façades, and sustainability reporting, accounting. Organ. Soc. **40**, 78–94 (2015)
49. Dhaliwal, D.S., Li, O.Z., Tsang, A., Yang, Y.G.: Corporate social responsibility disclosure and the cost of equity capital: the roles of stakeholder orientation and financial transparency. J. Accountability Public Policy **33**, 328–355 (2014)
50. Lundin, M.: Assessment of the environmental sustainability of urban water systems. Ph.D. thesis, Chalmers University of Technology (1999)
51. Larsen, T.A., Gujer, W.: The concept of sustainable urban water management. Water Sci. Technol. **35**, 3–10 (1997)
52. International Integrated Reporting Council (IIRC). Towards integrated reporting: communicating value in the 21st century. Technical report (2011)
53. OECD. Environmental performance reviews: Italy 2013. Technical report (2013)

New Paths for the Application of DCI in Social Sciences: Theoretical Issues Regarding an Empirical Analysis

Riccardo Righi[1](✉), Andrea Roli[2], Margherita Russo[1], Roberto Serra[3], and Marco Villani[3]

[1] Department of Economics, University of Modena and Reggio, Emilia, Italy
{riccardo.righi,margherita.russo}@unimore.it
[2] Department of Computer Science and Engineering,
University of Bologna, Bologna, Italy
andrea.roli@unibo.it
[3] Department of Physics, Informatics and Mathematics,
University of Modena and Reggio, Emilia, Italy
{rserra,marco.villani}@unimore.it

Abstract. Starting from the conceptualization of 'Cluster Index' (CI), Villani *et al.* [16,17] implemented the 'Dynamic Cluster Index' (DCI), an algorithm to perform the detection of subsets of agents characterized by patterns of activity that can be considered as integrated over time. DCI methodology makes possible to shift the attention into a new dimension of groups of agents (i.e. communities of agents): the presence of a common function characterizing their actions. In this paper we discuss the implications of the use in the domain of social sciences of this methodology, up to now mainly applied in natural sciences. Developing our considerations thanks to an empirical analysis, we discuss the theoretical implications of its application in such a different field.

1 Introduction

Taking advantage of two information theory concepts (integration and mutual information), Giulio Tononi introduced a new concept, the 'cluster index' (CI) [12–15], and demonstrated that neurons with integrated profiles of activity over time (i) have similar functions and (ii) have a location that is independent from anatomical proximity. Following this pioneering contribution, Villani and co-authors [16,17] developed an algorithm for the detection of subsets of agents introducing the comparison between the CI of an observed subset and the CI of a homogeneous system. The resulting algorithm, named by the authors 'Dynamic Cluster Index' (DCI), is able to produce a final ranking of all possible subsets that can be considered in any initial set. So far, DCI has been tested in research areas of artificial network models, of catalytic reaction networks and of biological gene regulatory systems [16,17], giving a contribution to the problem of identifying emergent meso-level structures [3].

© Springer International Publishing AG 2017
F. Rossi et al. (Eds.): WIVACE 2016, CCIS 708, pp. 42–52, 2017.
DOI: 10.1007/978-3-319-57711-1_4

The implementation of the DCI algorithm opens new paths for addressing socio-economic problems regarding the analysis of group of agents. In social sciences, the detection and the analysis of communities typically are performed through the consideration of similar characteristics of agents, or through the analysis of the observed network structure. Indeed, DCI methodology makes possible to shift the attention into a new dimension of organizations of agents: the presence of a common function characterizing their actions. Developing our considerations thanks to an empirical analysis, in this paper we discuss the theoretical implications of the use of this methodology in the domain of social sciences.

The paper is structured as follows. Section 2 proposes an overview of the theoretical elements of the CI proposed by Tononi *et al.* [12–15] and of the DCI as proposed by Villani *et al.* [16,17]. Section 3 presents the issue addressed in the case study: the evaluation of the network innovation regional policy implemented by Tuscany Region (Italy) in the programming period 2000–2006. Section 4 discusses the advantages of applying DCI in a context where the application of Complex Network modeling of community detection comes up against the absence of stepwise processes of formation/dissolution of relational structures. Section 5 focuses on the analytical problem of defining what the "activity" of an agent is. In Sect. 6, theoretical considerations regarding the application of DCI in a socio-economic context of analysis are illustrated. Section 7 underlines the potentiality of DCI analysis to investigate unobserved relations. Section 8 concludes summarizing the investigation of functional communities (group of agents that share a common function) in the landscape of community detection techniques, and highlights the potentialities of the application of DCI in socio-economic analyses aiming at detecting emerging functional communities.

2   DCI Analysis for the Detection of Functional Groups of Agents

Dynamic Cluster Index analysis (DCI) takes its origin from the neurological studies of Giulio Tononi in the 90's. Tononi supposed that neurons with similar functions show high level of coordination in their behaviors over time, independently from being, or not, situated in the same brain region[1]. Tononi introduced the notion of functional cluster, defining it as "a set of elements that are much more strongly interactive among themselves than with the rest of the system, whether or not the underlying anatomical connectivity is continuous" [12]. Thus, these functional clusters are made up of interactive neurons that produce an internal exchange of information (among neurons belonging to the same group) stronger

---

[1] In the field of neurological activity, two theories have always been opposed: the first, a localizationist theory sustains that the brain is divided into separate areas characterized by specific functions, while the second sustains the presence of a holistic scheme of the brain activity. Neither of these formulations were compatible with the hypothesis of the presence of groups of neurons that, regardless of their position, have specific and common functions.

than the exchange of information that the same neurons have with the rest of the system. The identification of these groups of neurons was realized by making use of two information theory concepts derived from the Shannon entropy: integration (I) and mutual information (MI). Taking advantage of these measurements a new concept was introduced [14]: the cluster index (CI). Formally, the CI of a subset X is written as follows

$$CI(X) = I(X)/MI(X, U \backslash X) \tag{1}$$

where X is the *j-th* subset of the whole system U, and is made up of k elements. Thanks to the CI, evidences of neurons that could be considered to belong to specific functional sub-systems, even if they do not participate in the same cerebral area, were found [14]. These studies demonstrated that neurons showing integrated profiles of activity over time (grouped together thanks to the analysis of CI) have (i) similar functions in the brain, and (ii) not necessarily show anatomical proximity.

Since integration and mutual information values depend on the size of the subsystem that is under analysis, in order to normalize them it is possible to make use of a so-called homogeneous system where the variables do not have correlation[2] [14,16,17]. Finally, the level of significance of the normalized CI (calculated as a statistical distance of the normalized CI, or CI', of the considered subset from the average CI of a subset having the same size extracted from the homogeneous system) is the value according to which the final ranking of all possible subsets is produced [14]:

$$CI'(X) = \frac{I(X)}{\langle I_h \rangle} \Big/ \frac{M(X, U \backslash X)}{\langle M_h \rangle} \tag{2}$$

$$t_{ci} = \frac{CI'(X) - \langle CI'_h \rangle}{\sigma(CI'_h)} \tag{3}$$

where $\langle I_h \rangle$ and $\langle M_h \rangle$ indicate respectively the average integration of subsets of dimension k belonging to the homogeneous system and the average mutual information of these subsets with the remaining part of the homogeneous system. $\langle CI'_h \rangle$ and $\sigma(CI'_h)$, respectively the mean and the standard deviation of normalized cluster indices of subsets that have the same size of X and that belong to the homogeneous system, are used to compute the statistical index t_{ci}.

Following these studies, Villani and co-authors borrowed the concept of CI and t_{ci} introducing it in research areas of artificial network models, of catalytic reaction networks and of biological gene regulatory systems, giving a contribution to the problem of identifying emergent meso-level structures [8,16,17]. Moreover, since an exhaustive computation of this statistic (t_{ci}) is possible only in small artificially designed networks, like those that were initially used to test

---

[2] A homogeneous system is a system having the same number of variables of the system to which it is referred; each variable has a random generated behavior in accordance with the probability of the states it assumes in the reference system.

the efficacy of the method [3,16,17], Villani and co-authors overcome the problem of computational duration in bigger initial set introducing a heuristic investigation in the algorithm [3]. The creation and the implementation of the DCI algorithm opened new path for analyses of community detection. The process of investigation made possible by DCI, allows researcher to look for groups characterized by levels of behavioral integration that, being significantly far from randomness, reveal the presence of at least one specific function jointly pursued by all the involved members. So far, the detection of groups (or communities) has been typically performed by focusing on similarity of agents' characteristics, or through the analysis of the observed network structure. With DCI methodology it is possible to shift the attention into a new dimension of organizations of agents. Since low levels of entropy are determined by the repetition of specific combinations of the statuses of a multiplicity of individuals over time, the emergence of a dynamic pattern unveils the alignment of the actions of these individuals towards a common function.

3 The Case Study: A Regional Programme to Support Innovation Networks

To interpret the application of DCI in social sciences, in this section we present an empirical analysis on a regional programme implemented by Tuscany Region (Italy) in the period 2000–06, aimed at supporting innovation networks. The programme sustained the development of innovation processes by fostering interactions between local agents (enterprises, universities, public research centers, local government institutions, service centers, etc.). The rationale of those policies is traced back in the need to overcome the difficulties of an industrial structure characterized by small and medium size enterprises in traditional sectors (such as textile, fashion, marble), generally not well linked to research activities. The various policy measures allowed the granting of funds exclusively to projects promoted by group of agents. Based on complex network analysis of innovation processes [5–7], that policy has been analysed by Caloffi, Rossi and Russo who highlighted the ontology of the programme [11], created an original data base with information on the agents participating to the programme and investigated the characteristics of the agents participating in the network-projects [1,2,9,10]. Since the goal of the policy programme was to favor collaborations in order to stimulate the flourishing of innovation, a key research question is to assess whether these policies were contributing to the formation of communities of innovative agents.

Started in 2002 (ending in 2008), the programme of public policies was composed of nine waves not uniformly distributed over time: they had different durations and they overlapped, producing periods in which no wave was active and periods in which three waves were simultaneously active. In addition, each wave presented specific features with regard to:

- the presence of constraints regarding the composition of the partnerships;
- the presence of constraints regarding the possibility of participating in more than one project in the context of the same wave;
- the technological domains in which projects were asked to operate;
- the amount of financial resources made available;
- the percentage (on the basis of the costs) of the grants of funds to each single project.

Another key feature was that agents could participate in more than one project (irrespective of the waves in which the projects were submitted) and they participate repeatedly with different partners. This means that every observed partnership could be made up of a different combination of agents and this element, added the those described above, increases the difficulties of grasping the network dynamics emerged over time. Looking at Fig. 1, that presents agents' participations over the nine waves (each agent keeps always the same coordinates across the nine graphs, and is represented only if in the correspondent wave it was active), it is possible to immediately capture how all the features described above produced a discontinuous evolution of the network. In such a picture, it seemed not appropriate to study the flourishing of communities looking at the stepwise creation of network frameworks. The degree of formation and of dissolution of connections was so high that brought to a situation of intense discontinuity over time.

4   DCI Analysis and Discontinue Network Dynamics

The peculiarities described in the previous paragraph strongly affected the possibility to study the detection of communities by taking into account the stepwise formation of relational structures among agents. Moreover, since the specific objective of our analysis is to investigate the presence of agents having a common function (and in this case study the common functions are likely to be related with the participation in projects and with the development of innovations), the analysis could not start from information on project-networks. In fact, following Tononi et al. [14], the presence of a functional groups is associated with the emergence of specific patterns of interactions, more than with the progressive establishment of relational community structures. We are not looking for the presence of a connective architecture that in some instant in time could reveal the presence of a community: we are looking for dynamics of interactions that reveal an alignment of agents and, consequently, that imply the joint pursuit of a common goal. Thus, rather than considering the formation/dissolution of the established connective structures, the idea was to focus the attention into how agents behaved over time. These considerations led to the formulation of a specific research question: *how is it possible to investigate the presence of functional communities, in a context in which the involved agents interact over time, but through networks' configurations that are discontinously changing over time?*

To answer this question, DCI analysis seemed to be appropriate. Any kind of relational information (topology of the project-network configuration) was

wave1_2002_ITT   **wave2_2002_171**   **wave3_2002_172**

wave4_2004_171   **wave5_2004_171A**   **wave6_2005_171**

wave7_2006_VIN   **wave8_2007_171**   **wave9_2008_171**

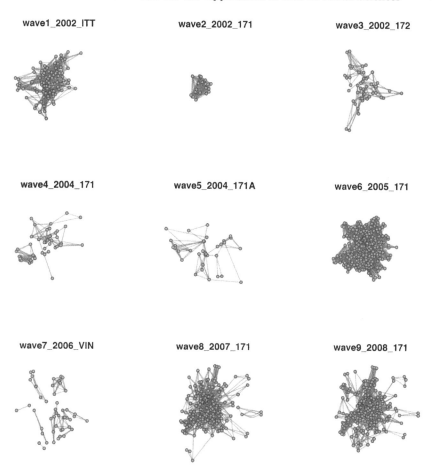

Fig. 1. Graphs representing the agents that took part in each of the nine waves. Each wave is labelled with a progressive number reflecting the overall chronological order, with the year in which it began, and with a code associated with the specific kind of the policies promoted. To every node is attributed the same position in all the graphs. Nodes are represented only if in the corresponding wave they participate in at least one project. The edges represent the common participations in at least one project (in the context of the corresponding wave). All the graphs are plotted on the same scale. Fruchterman Reingold layout. Source: our elaboration on Region Tuscany Network Policies (2000–06). Figure generated using R (igraph package).

abandoned and the attention was centered on the integration of agents' behaviors over time. The idea was to observe what agents did over the period of the considered public policies.

5 The Application of DCI in a Socio-economic Context of Analysis

While up to now the motivations that led to the application of DCI analysis in a context of policy evaluation have been described, there are some theoretical implications that need to be taken into account. Moving to a socio-economic context of analysis necessarily involves three different considerations. First of all, it is more difficult to isolate the system under analysis, and since agents may be influenced by external elements, a high degree of openness can compromise the analysis enhancing the difficulty to investigate the dynamic of agents' subsystems that are the focus of the analysis. Secondly, while in the context of artificial and chemical systems initial conditions can be easily established, making it relatively easy to understand the relation between them and the behaviors of agents, this is much harder to do in socio-economic systems. Finally, there is also an ontological issue that must be considered: interactions of socio-economic agents may reveal behavioral patterns, but agents do not necessarily follow deterministic rules of interaction.

Without having the purpose to investigate the specific and detailed functions of subsets of agents, the aim of applying DCI in social sciences is to evaluate the presence of common objectives which, if they exist, are likely to imply the presence of information embodied in integrated activities. To have a shared function (as it is supposed that members of a community have) implies that involved agents behave with some kind of physical order that help them to reach together their common aim. The contraposition is between random behaviors, which do not give sense to a group action, and the physical order through which agents act to perform a shared function [4]. The information that the integration of behaviors contains is exactly what DCI takes into account.

6 The Data Structure and the Definition of Agents' Activity

The most important aspect that we had to face in order to apply DCI analysis to the case study regarded the definition of the agents' activity statuses. Since the available information regarded the starting and the ending date of agents' participations in the projects, we defined that each agent had to be considered active in those moments in which it was participating in at least one project. Working on these dates, it was possible to define a complete behavioral profile for the agents involved in the six years of the policy programme: in every instant it was possible to observe which agents were active and which were not. First of all boolean variables were defined, in accordance with the participation of agents in at least one project. Moreover, since information about the number of projects in which every agent was active in each single instant was available, it was possible to define a second series of variables describing the variation of the levels of activity. These variables assume four different values that correspond to four different situations:

- the agent is not participating in any project (no activity);
- the agent is participating in a number of projects that is higher than the number of projects in which it was participating in the previous instant (increasing activity);
- the agent is participating in a number of projects that is equal to the number of projects in which it was participating in the previous instant (constant activity);
- the agent is participating in a number of projects that is lower than the number of projects in which it was participating in the previous instant (decreasing activity).

With this second series of variables, we create of a model that takes into account a second order Markov condition. Agents' activity is not described for what is in each instant, but for what is in the present conditioned by what was in its nearest past. This more detailed description of the agents' behavioral dynamics allows evaluating entropy measurements with respect to how behaviors were changing. By doing so, more information is introduced in the model.

To conclude this section, it is important to remark the fact that, even if information regarding participation in projects was the only one available, it was the best we could ask for. As it would be clearer in the next section, policies contributed in the rising of other kinds of activities and other kinds of interactions among agents. Nevertheless, what the measures of policy fostered them to do was to design project-networks. Thus, the participation in projects, with regard to the specific context of analysis, has to be considered the ultimate kind of activity to which they tended, and so the most relevant among all.

7   DCI Analysis and Unobserved Relations Among Agents

Finally, moving to the conclusion of this contribution, a last theoretical point regarding the application of DCI in such a context of analysis has to be introduced. As it has been described above, DCI analysis allows researchers to investigate the presence of communities of agents without considering any kind of information about the topology of the network, a specific feature that assumes a particular meaning in the case study presented here. Since the considered policy measures financed exclusively project-networks, an immediate solution to a problem of community detection could be to focus on the structure of common participations. Of course, in the analysis of the considered public policy these relations represent the crucial types of interactions but, nevertheless, they represent only one type among all those that during the entire policy programme could have occurred. Thus, a crucial question arises: *is it not possible that functional groups could have bloomed not only through common participations in projects?*

The answer is affirmative. It is likely to admit that during the considered period of time agents did not have relations among themselves exclusively on the basis of common participations in projects and, consequently, it is not

unreasonable to think that other unobserved relations could have been important in agents' decisions about final participations in regional public policies.
E-mails, phone calls, work meetings, conferences, projects different from the
ones observed, standard working relations, trading operations, etc.: these are
some possible examples of relations that could have had a role in generating
communities. Obviously, from the point of view of our analysis, it would have
been incredibly interesting to consider and to study these relations, but they
remained unobserved. To focus the attention on agents' participations in the
policies means to consider their ultimate activity: that kind of activity that is
the final result of all other interactions that occurred. Thus, it is possible to
assert that the application of DCI methodology, even if without taking into
account network information, opens the opportunity to embed in the result the
whole set of relations that were active during the period under observation.

8   Conclusion

DCI analysis shows several theoretical aspects that encouraged its application
in the domain of social sciences. It has been clarified that the motivation for
which it was developed opens news possibilities for researchers: the investigation
of functional communities (group of agents whose patterns of activity, being far
from randomness, reveal the presence of a common function) is something new
in the landscape of community detection methodologies. It has been described
that in a context of analysis in which available relational data show a strong
discontinuity over time, it would have been difficult to look at the stepwise
formation of specific network structures, while DCI does not require this kind of
condition. It has been remarked that, in a socio-economic analysis, the intrinsic
nature of the context and of the agents considered do not affect the possibility
to apply DCI to investigate the presence of integrated behaviors. Then, some
considerations about the definition of the activity status have been made. In
particular, a model that takes into account a second order Markov condition has
been described, and the importance of the kind of activity that are considered
was explained. Finally, it has been highlighted that DCI allows researchers to
make considerations upon the complete set of relations that occurred in the
considered period, without necessarily constraining the analysis to those that
were observable. This final point is crucial since relations that remain unobserved
could have had a role in the process of the emerging of functional communities.

   With this work we explained the theoretical considerations that open the
possibility to apply DCI analysis in socio-economic contexts. It is important to
remark that depending on the specific application (and on the availability of
data), the results of the DCI should be integrated with the results of the application of other kinds of community detection algorithms and with analyses of
the network structure. In particular, the comparison between network communities and DCI functional groups should give helpful insights in socio-economic
complex systems.

References

1. Caloffi, A., Rossi, F., Russo, M.: The emergence of intermediary organizations: a network-based approach to the design of innovation policies. In: Handbook On Complexity And Public Policy, pp. 314–331. Edward Elgar Publishing, Cheltenham GBR (2015). ISBN: 978-1-78254-951-2
2. Caloffi, A., Rossi, F., Russo, M.: What makes SMEs more likely to collaborate? Analysing the role of regional innovation policy. Eur. Plan. Stud. **23**(7), 1245–1264 (2015). http://dx.doi.org/10.1080/09654313.2014.919250
3. Filisetti, A., Villani, M., Roli, A., Fiorucci, M., Serra, R.: Exploring the organisation of complex systems through the dynamical interactions among their relevant subsets. In: Proceedings of the European Conference on Articial Life 2015 (ECAL 2015), vol. 13, pp. 286–293 (2016). https://mitpress.mit.edu/sites/default/files/titles/content/ecal2015/ch054.html
4. Hidalgo, C.: Why Information Grows: The Evolution of Order, from Atoms to Economies. Basic Books, New York (2015)
5. Lane, D.A.: Complexity and innovation dynamics. In: Handbook on the Economic Complexity of Technological Change. Edward Elgar Publishing, Cheltenham (2011). ISBN: 978-0-85793-037-8
6. Lane, D.A., Maxfield, R.R.: Foresight, complexity, and strategy. In: The Economy as an Evolving Complex System II. Westview Press (1997). ISBN: 978-0-201-32823-3
7. Lane, D.A., Maxfield, R.R.: Ontological uncertainty and innovation. J. Evol. Econ. **15**(1), 3–50 (2005). doi:10.1007/s00191-004-0227-7
8. Roli, A., Villani, M., Caprari, R., Serra, R.: Identifying critical states through the relevance index. Entropy **19**(2), 73 (2017). Special issue: Complexity, Criticality and Computation, MDPI AG, Basel, Switzerland
9. Rossi, F., Caloffi, A., Russo, M.: Networked by design: can policy requirements influence organisations' networking behaviour? Technol. Forecast. Soc. Change **105**, 203–214 (2016). doi:10.1016/j.techfore.2016.01.004
10. Russo, M., Caloffi, A., Rossi, F.: Evaluating the performance of innovation intermediaries: insights from the experience of Tuscany's innovation poles. In: Plattform Forschungs- Und Technologieevaluierung, vol. 41, pp. 15–25 (2015). ISSN: 1726–6629
11. Russo, M., Rossi, F.: Cooperation networks and innovation. A complex system perspective to the analysis and evaluation of a EU regional innovation policy programme. Evaluation **15**, 75–100 (2009). doi:10.1177/1356389008097872
12. Tononi, G., McIntosh, A.R., Russell, D.P., Edelman, G.M.: Functional clustering: identifying strongly interactive brain regions in neuroimaging data. NeuroImage **7**(2), 133–149 (1998). doi:10.1006/nimg.1997.0313
13. Tononi, G., Sporns, O., Edelman, G.M.: A complexity measure for selective matching of signals by the brain. Proc. Natl. Acad. Sci. USA **93**(8), 3422–3427 (1996). ISSN: 0027–8424
14. Tononi, G., Sporns, O., Edelman, G.M.: A measure for brain complexity: relating functional segregation and integration in the nervous system. Proc. Natl. Acad. Sci. USA **91**(11), 5033–5037 (1994). http://www.jstor.org/stable/2364906. ISSN: 0027–8424
15. Tononi, G., Sporns, O., Edelman, G.M.: Measures of degeneracy and redundancy in biological networks. Proc. Natl. Acad. Sci. **96**(6), 3257–3262 (1999). http://www.pnas.org/content/96/6/3257. ISSN: 0027–8424,1091–6490

16. Villani, M., Benedettini, S., Roli, A., Lane, D., Poli, I., Serra, R.: Identifying emergent dynamical structures in network models. In: Bassis, S., Esposito, A., Morabito, F.C. (eds.) Recent Advances of Neural Network Models and Applications. Smart Innovation, Systems and Technologies, vol. 26, pp. 3–13. Springer, Cham (2014). doi:10.1007/978-3-319-04129-2_1

17. Villani, M., Filisetti, A., Benedettini, S., Roli, A., Lane, D., Serra, R.: The detection of intermediate-level emergent structures and patterns. In: ECAL, pp. 372–378 (2013)

MapReduce in Computational Biology - A Synopsis

Giuseppe Cattaneo[1], Raffaele Giancarlo[2], Stefano Piotto[3] (iD),
Umberto Ferraro Petrillo[4], Gianluca Roscigno[1(✉)], and Luigi Di Biasi[3]

[1] Dipartimento di Informatica, Università degli Studi di Salerno,
84084 Fisciano, SA, Italy
{cattaneo,giroscigno}@unisa.it
[2] Dipartimento di Matematica ed Informatica,
Università di Palermo, 90133 Palermo, PA, Italy
raffaele.giancarlo@unipa.it
[3] Dipartimento di Farmacia, Università degli Studi di Salerno,
84084 Fisciano, SA, Italy
piotto@unisa.it
[4] Dipartimento di Scienze Statistiche,
Università di Roma "La Sapienza", 00185 Roma, Italy
umberto.ferraro@uniroma1.it

Abstract. In the past 20 years, the Life Sciences have witnessed a paradigm shift in the way research is performed. Indeed, the computational part of biological and clinical studies has become central or is becoming so. Correspondingly, the amount of data that one needs to process, compare and analyze, has experienced an exponential growth. As a consequence, High Performance Computing (HPC, for short) is being used intensively, in particular in terms of multi-core architectures. However, recently and thanks to the advances in the processing of other scientific and commercial data, Distributed Computing is also being considered for Bioinformatics applications. In particular, the MapReduce paradigm, together with the main middleware supporting it, i.e., Hadoop and Spark, is becoming increasingly popular.

Here we provide a short review in which the state of the art of MapReduce bioinformatics applications is presented, together with a qualitative evaluation of each of the software systems that have been here included. In order to make the paper self-contained, computer architectural and middleware issues are also briefly presented.

Keywords: Bioinformatics · Distributed computing · MapReduce · Hadoop · Spark

1 Introduction

The development of sequencing technologies has caused a stunning reduction in sequencing costs, well illustrated in Fig. 1, whose effect is an exponential increase

© Springer International Publishing AG 2017
F. Rossi et al. (Eds.): WIVACE 2016, CCIS 708, pp. 53–64, 2017.
DOI: 10.1007/978-3-319-57711-1_5

in the production of genomic and proteomic sequences that need to be analyzed. Unfortunately, computer hardware costs have not kept the same reduction pace, causing an economic problem to research areas that use sequencing, i.e., the entire Life Sciences. Those aspects are well summarized in [1,2], where it is envisioned that the cost of analyzing sequence data may be a factor of 100 times more than the cost of its production. Solutions are also proposed, one of which is to use Cloud and Distributed Computer Systems to support the computer-related aspects of research in the Life Sciences. Although *High Performance Computing* (HPC) is a fundamental part of Bioinformatics and Computational Biology, it has been mainly used for computationally hard problems such as prediction of protein structure. Unfortunately, the scenario has changed dramatically due to the simple sheer quantity of data to be processed: even the simple task of counting the number of k-mers in a set of strings, a fundamental problem in Genomics and Epigenomics (see, e.g., [3–5]), is being addressed with the use of multi-processor architectures, e.g., [6]. As opposed to multi-processor architectures, the use of Distributed Architectures is emerging only now and it seems to be the future, due to the scalability that this type of architecture grants. In order to foster further development of Bioinformatics and Computational Biology distributed software systems and platforms, here a short review of this area is provided. Moreover, we also identify some desirable properties that those software systems should have and provide a corresponding evaluation of them. Specifically, Sect. 2 is dedicated to a short description of distributed computing in the management of Big Data, with emphasis on the architectural and middleware issues. In Sect. 3, we focus on MapReduce implementations in bioinformatics. Finally, in Sect. 4, some conclusions are offered.

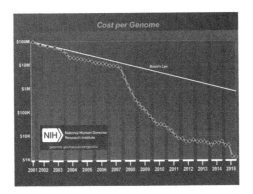

Fig. 1. Cost (on a logarithmic scale) to obtain a good draft of the Human Genome, as sequencing technology improves over the years. The well known Moore's law governing computer hardware costs is also shown. In 2008, it became cheaper to produce a DNA sequence with respect to the hardware that would be needed to analyze and store it. The figure is taken from [7]

2   Distributed Computing in the Management of Big Data: Architectural and Middleware Issues

Many bioinformatics problems can be solved by partitioning the input in a number of independent parts that can be processed independently. Therefore, many time-consuming applications in bioinformatics have been expressly designed to exploit parallelism of the multi-core shared-memory architectures. Due to a number of architectural bottlenecks mainly related to the need of multiple cores to share the same memory or I/O buses, multi-core shared-memory architectures do not allow to efficiently use more than a rather limited number of cores. For this reason, the performance of a parallel algorithm developed for those architectures may not scale when the number of cores goes beyond a certain threshold. An application can be considered *scalable* when its execution time is proportionally reduced when the number of computing units is increased.

The above shortcoming of multi-core shared-memory architectures is bypassed by resorting to *Distributed Systems*. It is well-know that this term refers to a collection of independent computers (also called *nodes*) that communicate over a network. The interested reader can find an introduction in this subject in [8]. Theoretically, with this approach, an arbitrary large number of nodes can be added to the system by exploiting its inherent scalability (i.e., *scale out*). This is possible since each node can access local resources without competing with other nodes. This feature can radically improve the performance of any properly designed application reducing its execution time.

As a drawback, the adoption of a distributed approach often requires more specific and complex skills to design and develop a distributed algorithm, given the target architecture. To promote the transition toward distributed systems, many new programming paradigms have been proposed in recent years. Among those, MapReduce [9] paradigm is becoming a *de facto* standard. It is described next, together with the two main middleware systems supporting it, i.e., Apache Hadoop [10] and Apache Spark [11,12].

2.1   MapReduce Paradigm

It is based on the proper definition of two functions: *map* and *reduce*. Assuming that the input is organized as a set of *<key, value>* pairs, the generic map function takes as input one of these pairs and returns, as output, a set of intermediate *<key, value>* pairs. The reduce function is then used to process all the intermediate pairs having the same *key*, typically returning a synoptic rendering of the input. Map and reduce functions are executed, as tasks, by workers running on the nodes of a distributed system complying to a MapReduce framework. The communication between workers running map functions and workers running reduce functions is accomplished in an automatic and transparent way by the underlying framework (*implicit parallelism*). Therefore, the programmer can focus on the definition of the map and reduce functions, since all the aspects related to the execution in a distributed setting (e.g., the number of concurrent map and reduce tasks to issue) are addressed via the proper definition of configuration variables.

An overview of MapReduce paradigm is depicted in Fig. 2.

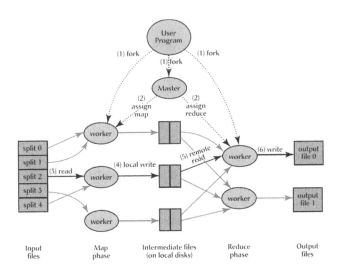

Fig. 2. Workflow of a MapReduce application (taken from [9]). Input files are split in several parts, with each part processed by a distinct map task according to a user-defined function. Each of these tasks emits, as output, 0, 1 or more intermediate $<K,V>$ pairs. Then, these pairs are shuffled and sorted so that all the pairs having the same key are processed by the user-defined function of a same reduce task. These produce, as output, a set of $<K,V>$ pairs that are appended to a different part of the same logical file.

Apache Hadoop. It is currently the most popular and mature framework supporting MapReduce. It is mainly composed of two modules: YARN (*Yet Another Resource Negotiator*) [13] and HDFS (*Hadoop Distributed File System*) [14]. YARN is a data processing framework supporting the execution of distributed algorithms through different types of computing paradigms. HDFS is a distributed and block-structured file system designed to run on commodity hardware and able to provide fault tolerance through replication of data.

In general, a Hadoop cluster consists of a single *master node* and multiple *slave nodes*: the master node runs the `Resource Manager` and the `Name Node` services, while slave nodes run the `Data Node` and the `Node Manager` services. On the master node, the `Resource Manager` arbitrates the assignment of computational resources to applications. On the slave nodes, the `Node Manager` monitors and keeps informed the `Resource Manager` about the status of the node. Again, on the master node, the `Name Node` service maintains the directory tree of all files existing in the HDFS and keeps tracks of where data blocks are physically placed. On the slave nodes, the `Data Node` service maintains a subset of the HDFS data blocks using the local storage.

Applications are run on Hadoop via an `Application Master`. This is a service instantiated by a `Node Manager` on a slave node upon a request coming from the `Resource Manager`. Once created, it asks the Hadoop framework for all the resources required to perform a computation (mainly in terms of CPU and memory). The `Resource Manager` responds by reserving to the application a set of *workers* (also called `Containers`) running on one or more slave nodes (a worker is the basic processing unit in Hadoop to execute map or reduce tasks).

HDFS: Hadoop Distributed File System. In a distributed system it is often more efficient to run a task on local data, rather than to move the data where the task is running. In fact, one of the main characteristics of Hadoop is its ability to exploit data local computing. In particular, HDFS provides functionality to enable applications to move closer to the data, minimizing network congestion and increasing the overall throughput of the system.

It is a distributed and block-structured file system, optimized to run also on commodity hardware and able to provide fault tolerance through replication of data. The HDFS is able to reliably maintain very large files, in fact, it works by automatically splitting large files in smaller blocks and spreading them across nodes in a network.

Lifetime of a Hadoop MapReduce Algorithm. The execution of a Hadoop MapReduce application takes place in two consecutive (and potentially overlapping) phases: the map phase and the reduce phase. During the map phase, one or more map tasks are run by workers on the slave nodes of the Hadoop cluster. Each worker may run one task a time, while several workers may run in parallel on the same slave node. When an application is running, a worker in the cluster is dedicated to the `Application Master` service.

The execution of a map task goes through four phases. At startup, the task initializes the data structures required for managing the input and the output of the task (*init* phase). Then, each worker begins the execution of the map functions (*execution* phase). As soon as output pairs are returned, these are saved in a temporary memory buffer. When the buffer gets almost full or when the map functions execution ends, the output pairs are sorted, partitioned according to the destination reduce task and written on disk (*spilling* phase). Then, data belonging to each partition are moved to the slave nodes where the reduce tasks will process them.

The execution of a reduce task requires three phases. At the beginning, all the pairs produced by map tasks and included in a certain partition are moved on the node where the reduce tasks assigned to that partition will be run (*shuffle* phase). As soon as new pairs are received by a node, they are sorted in order to keep them grouped according to their key (*sort* phase). Finally, for each group of pairs with the same key, a reduce function will be run by a worker on that node.

Apache Spark. It is a distributed framework that supports applications with acyclic data-flow model and in-memory computing. This framework also preserves the scalability and fault tolerance of MapReduce-compliant frameworks. Spark provides APIs in Scala, Java and Python programming languages, and it can be used for programs that reuse a working set of data across multiple parallel operations, such as iterative algorithms.

Apache Spark also has a *master/slaves* architecture like Hadoop. In particular, there are two main services: `Worker` and `Driver`. A `Worker` is a service running on any slave node that can execute application code. An `Executor` is a process launched for an application on a `Worker` node that runs tasks and keeps data in memory or disk storage. In addition, each application has its own `Executors`.

The `Driver` service runs on the master node and it manages the `Executors` through a cluster resource manager, e.g., stand-alone cluster manager of Spark, Apache Mesos or Hadoop YARN. In particular, the `Workers` create `Executors` for the `Driver`, and then it can run tasks in those `Executors`.

The main feature of Spark is the *Resilient Distributed Dataset* (RDD). It represents a read-only collection of objects partitioned across a set of nodes that can be rebuilt if a partition is lost. An RDD can be used to cache data in memory across nodes and it can be reused in multiple MapReduce-based operations. Several parallel operations can be performed on RDDs, such as: *reduce*, *collect* and *foreach*. The *reduce* operation combines dataset elements using an associative function to produce a result, while the *collect* operation sends all elements of the dataset to the `Driver`. Instead, the *foreach* operation passes each element through a user provided function.

In addition, Spark also provides `Datasets` and `DataFrames` features. The first is a distributed collection of data providing an interface that combines the benefits of RDDs with those of Spark SQL's optimized execution engine. A `Dataset` can be constructed from objects and then manipulated using functional transformations, such as: *map*, *flatMap*, *filter*, etc. Instead, a Spark `DataFrame` is a `Dataset` organized into named columns, and it is equivalent to a table concept in a relational database, but providing more optimizations. Roughly speaking, a `DataFrame` is represented by a `Dataset` of rows.

Apache Spark can have the Hadoop framework as the underlining middleware. Moreover, as opposed to Hadoop, it is not limited to support the MapReduce paradigm. Finally, it allows the combination of streaming and batch processing, while Hadoop can be only used for batch applications.

3   State of Art of MapReduce-Based Software in Bioinformatics

In this section, we report a classification of bioinformatics software tools that are based on the MapReduce paradigm. The vast majority of them is supported by Hadoop and some by Spark. A new special purpose framework has also been proposed in [15], i.e., GATK.

As far as the qualitative evaluation of the presented software, the properties that are desirable each must have are the following.

(a) **Programming Platform (PP).** This property means that the software has been designed as a flexible programming platform to be used for a set of customizable analyses. This property is held by any application that allows custom queries or high level programming interface (API) to develop new applications. For example the Hadoop-based BioPig [16] tool is considered a good platform to solve bioinformatics problems and, therefore, holds this property.

(b) **Booster.** In some specific cases and in order to gain in time performance, MapReduce applications are developed by distributing multiple instances of a sequential package/library already existing, which is used as black box. Therefore, such a wrapping provides a boosting of the sequential package. For instance, CloudBLAST [17] has been developed starting from a well-known existing application (namely BLAST [18]) through a "wrapper" for the Hadoop framework. Another example is GRIMD [19] that was built on Microsoft Windows Server System and on Microsoft VPN PT2P protocol to set up secure connections between clients and server. GRIMD is fully configurable and permits the deploy of quantum mechanical [20] as well as molecular dynamics calculations [21,22] and genome analysis.

(c) **New Algorithm (NewA).** In this case the MapReduce application is developed based on a new algorithm (expressly designed for a target distributed architecture) to solve a classic or new bioinformatics problem. For instance, CloudAligner [23] has been specifically designed according to the MapReduce paradigm and then implemented in Hadoop.

(d) **New Algorithm and Engineering (NewAE).** This property holds if the MapReduce application is developed adopting an algorithm engineering approach [24,25] which, in addition to property (c), is supposed to provide highly tuned code.

The software can be divided into the following categories.

- *Tools and programming environments for the development of Bioinformatics applications.* We have included in this category programming tools and environments used to develop MapReduce pipelines and programs for bioinformatics (see Table 1), e.g., processing of HTS sequence data. In addition, we have also included libraries that provide interfaces that allow high level applications to work on files that have a standard format in bioinformatics, such as Hadoop-BAM [26].
- *Algorithms for Single Nucleotide Polymorphism Identification.* We have included in this category software that performs SNP identification and analysis (see Table 2).
- *Gene Expression Analysis.* We have included in this category software for gene expression analysis (see Table 3), e.g., gene set analysis for biomarker identification.

- *Sequence Comparison.* We have included in this category software for sequence comparison based on alignments and alignment-free methods (see Table 4).
- *Genome Assembly.* We have included in this category software for *de novo* genome assembly from short sequencing reads (see Table 5).
- *Sequencing Reads Mapping.* We have included in this category software for mapping short reads to reference genomes (see Table 6).
- *Additional Applications.* We have included in this category MapReduce bioinformatics applications for which there is only one implementation available (see Table 7).

Table 1. *Tools and programming environments for the development of Bioinformatics applications.* With reference to the main text, the software reported in this category is classified as shown. YES indicates that the property is held, while NO indicates otherwise. Other possible values indicate to what extent the property is held: GE stands for "to a great extent", ME stands for "to a moderate extent" and SE stands for "to a small extent".

Software	PP	Booster	NewA	NewAE
BioPig [16]	YES	YES	GE	NO
Cloudgene [27]	YES	NO	NO	NO
FASTdoop [28]	YES	NO	GE	GE
GATK [15]	YES	NO	GE	NO
Hadoop-BAM [26]	YES	NO	GE	NO
SeqPig [29]	YES	YES	GE	NO
SparkSeq [30]	YES	NO	SE	SE

Table 2. *Algorithms for Single Nucleotide Polymorphism Identification.* This table is analogous to Table 1

Software	PP	Booster	NewA	NewAE
BlueSNP [31]	YES	YES	NO	NO
Crossbow [32]	NO	YES	SE	NO

Table 3. *Gene Expression Analysis.* This table is analogous to Table 1

Software	PP	Booster	NewA	NewAE
Eoulsan [33]	NO	YES	NO	NO
FX [34]	NO	NO	GE	NO
MyRNA [35]	YES	YES	SE	SE
YunBe [36]	NO	NO	GE	NO

Table 4. *Sequence Comparison.* This table is analogous to Table 1

Software	PP	Booster	NewA	NewAE
Almeida et al. [37]	NO	NO	GE	NO
CloudBLAST [17]	NO	YES	NO	NO
HAFS [38]	YES	NO	GE	GE
K-mulus [39]	NO	YES	NO	NO
Nephele [40]	NO	NO	NO	NO
Strand [41]	NO	NO	GE	NO

Table 5. *Genome Assembly.* This table is analogous to Table 1

Software	PP	Booster	NewA	NewAE
CloudBrush [42]	NO	NO	GE	NO
Contrail [43]	NO	YES	GE	NO

Table 6. *Sequencing Reads Mapping.* This table is analogous to Table 1

Software	PP	Booster	NewA	NewAE
BlastReduce [44]	NO	NO	GE	NO
CloudAligner [23]	NO	NO	GE	NO
CloudBurst [45]	NO	NO	GE	NO
SEAL [46]	NO	YES	NO	NO
SparkSW [47]	NO	NO	GE	ME

Table 7. *Additional Applications.* This table is analogous to Table 1

Software	PP	Booster	NewA	NewAE
BioDoop [48]	NO	YES	SE	SE
Codon Counting [49]	NO	NO	GE	NO
GRIMD [19]	YES	YES	ME	ME
MrMC-MinH [50]	NO	NO	GE	NO
MrsRF [51]	NO	NO	GE	ME
PeakRanger [52]	NO	NO	GE	NO

4 Conclusion

We have presented the state of the art regarding the use of Distributed Computing, in particular software designed with MapReduce paradigm, in Bioinformatics and Computational Biology. Although such a body of work will grow in the future, it would be appropriate to follow good design and algorithm engineering approaches in order to use in full the available hardware and the scalability MapReduce offers.

References

1. Kahn, S.D.: On the future of genomic data. Science **331**, 728–729 (2011)
2. Mardis, E.R.: The $1,000 genome, the $100,000 analysis? Genome Med. **2**, 1–3 (2010)
3. Compeau, P.E.C., Pevzner, P.A., Tesler, G.: How to apply de Bruijn graphs to genome assembly. Nat. Biotechnol. **29**, 987–991 (2011)
4. Giancarlo, R., Rombo, S.E., Utro, F.: Epigenomic k-mer dictionaries: shedding light on how sequence composition influences in vivo nucleosome positioning. Bioinformatics **31**(18), 2939–2946 (2015)
5. Utro, F., Di Benedetto, V., Corona, D.F., Giancarlo, R.: The intrinsic combinatorial organization and information theoretic content of a sequence are correlated to the DNA encoded nucleosome organization of eukaryotic genomes. Bioinformatics **32**(6), 835–842 (2015)
6. Deorowicz, S., Kokot, M., Grabowski, S., Debudaj-Grabysz, A.: KMC 2: fast and resource-frugal k-mer counting. Bioinformatics **31**(10), 1569–1576 (2015)
7. National Human Genome Research Institute (NIH): The cost of sequencing a human genome (2016). https://www.genome.gov/sequencingcosts/
8. Tanenbaum, A.S., Van Steen, M.: Distributed Systems. Prentice-Hall, Upper Saddle River (2007)
9. Dean, J., Ghemawat, S.: MapReduce: simplified data processing on large clusters. Commun. ACM **51**, 107–113 (2008)
10. Apache Software Foundation: Hadoop (2016). http://hadoop.apache.org/
11. Apache Software Foundation: Spark (2016). http://spark.apache.org/
12. Zaharia, M., Chowdhury, M., Franklin, M.J., Shenker, S., Stoica, I.: Spark: cluster computing with working sets. In: Proceedings of the 2nd USENIX Conference on Hot Topics in Cloud Computing, vol. 10, pp. 1–7 (2010)
13. Vavilapalli, V.K., Murthy, A.C., Douglas, C., Agarwal, S., Konar, M., Evans, R., Graves, T., Lowe, J., Shah, H., Seth, S., et al.: Apache Hadoop YARN: yet another resource negotiator. In: Proceedings of the 4th annual Symposium on Cloud Computing, pp. 1–16. ACM (2013)
14. Shvachko, K., Kuang, H., Radia, S., Chansler, R.: The Hadoop distributed file system. In: IEEE 26th Symposium on Mass Storage Systems and Technologies (MSST), pp. 1–10. IEEE Computer Society, Washington, DC (2010)
15. McKenna, A., Hanna, M., Banks, E., Sivachenko, A., Cibulskis, K., Kernytsky, A., Garimella, K., Altshuler, D., Gabriel, S., Daly, M., et al.: The genome analysis toolkit: a MapReduce framework for analyzing next-generation DNA sequencing data. Genome Res. **20**, 1297–1303 (2010)
16. Nordberg, H., Bhatia, K., Wang, K., Wang, Z.: BioPig: a Hadoop-based analytic toolkit for large-scale sequence data. Bioinformatics **29**, 3014–3019 (2013)
17. Matsunaga, A., Tsugawa, M., Fortes, J.: CloudBLAST: combining MapReduce and virtualization on distributed resources for bioinformatics applications. In: IEEE Fourth International Conference on eScience. eScience 2008, pp. 222–229. IEEE (2008)
18. Altschul, S.F., Gish, W., Miller, W., Myers, E.W., Lipman, D.J.: Basic local alignment search tool. J. Mol. Biol. **215**, 403–410 (1990)
19. Piotto, S., Di Biasi, L., Concilio, S., Castiglione, A., Cattaneo, G.: GRIMD: distributed computing for chemists and biologists. Bioinformation **10**, 43–47 (2014)

20. Lopez, D.H., Fiol-deRoque, M.A., Noguera-Salvà, M.A., Terés, S., Campana, F., Piotto, S., Castro, J.A., Mohaibes, R.J., Escribá, P.V., Busquets, X.: 2-Hydroxy arachidonic acid: a new non-steroidal anti-inflammatory drug. PloS ONE **8**, 1–10 (2013)
21. Piotto, S., Concilio, S., Bianchino, E., Iannelli, P., López, D.J., Terés, S., Ibarguren, M., Barceló-Coblijn, G., Martin, M.L., Guardiola-Serrano, F., Alonso-Sande, M., Funari, S.S., Busquets, X., Escribá, P.V.: Differential effect of 2-hydroxyoleic acid enantiomers on protein (sphingomyelin synthase) and lipid (membrane) targets. Biochimica et Biophysica Acta (BBA)-Biomembranes **1838**, 1628–1637 (2014)
22. Piotto, S., Trapani, A., Bianchino, E., Ibarguren, M., López, D.J., Busquets, X., Concilio, S.: The effect of hydroxylated fatty acid-containing phospholipids in the remodeling of lipid membranes. Biochimica et Biophysica Acta (BBA)-Biomembranes **1838**, 1509–1517 (2014)
23. Nguyen, T., Shi, W., Ruden, D.: CloudAligner: a fast and full-featured MapReduce based tool for sequence mapping. BMC Res. Notes **4**, 171 (2011)
24. Cattaneo, G., Italiano, G.F.: Algorithm engineering. ACM Comput. Surv. (CSUR) **31**, 582–585 (1999)
25. Demetrescu, C., Finocchi, I., Italiano, G.F.: Algorithm engineering. Bull. EATCS **79**, 48–63 (2003)
26. Niemenmaa, M., Kallio, A., Schumacher, A., Klemelä, P., Korpelainen, E., Heljanko, K.: Hadoop-BAM: directly manipulating next generation sequencing data in the Cloud. Bioinformatics **28**, 876–877 (2012)
27. Schönherr, S., Forer, L., Weißensteiner, H., Kronenberg, F., Specht, G., Kloss-Brandstätter, A.: Cloudgene: a graphical execution platform for MapReduce programs on private and public clouds. BMC Bioinform. **13**, 1–9 (2012)
28. Ferraro Petrillo, U., Roscigno, G., Cattaneo, G., Giancarlo, R.: FASTdoop: a versatile and efficient library for the input of FASTA and FASTQ files for MapReduce Hadoop bioinformatics applications. Bioinformatics (2017). https://dx.doi.org/10.1093/bioinformatics/btx010
29. Schumacher, A., Pireddu, L., Niemenmaa, M., Kallio, A., Korpelainen, E., Zanetti, G., Heljanko, K.: SeqPig: simple and scalable scripting for large sequencing data sets in Hadoop. Bioinformatics **30**, 119–120 (2014)
30. Wiewiórka, M.S., Messina, A., Pacholewska, A., Maffioletti, S., Gawrysiak, P., Okoniewski, M.J.: SparkSeq: fast, scalable, Cloud-ready tool for the interactive genomic data analysis with nucleotide precision. Bioinformatics **30**, 2652–2653 (2014)
31. Huang, H., Tata, S., Prill, R.J.: BlueSNP: R package for highly scalable genome-wide association studies using Hadoop clusters. Bioinformatics **29**, 135–136 (2013)
32. Langmead, B., Schatz, M.C., Lin, J., Pop, M., Salzberg, S.L.: Searching for SNPs with Cloud computing. Genome Biol. **10**, 1–10 (2009)
33. Jourdren, L., Bernard, M., Dillies, M.A., Le Crom, S.: Eoulsan: a Cloud computing-based framework facilitating high throughput sequencing analyses. Bioinformatics **28**, 1542–1543 (2012)
34. Hong, D., Rhie, A., Park, S.S., Lee, J., Ju, Y.S., Kim, S., Yu, S.B., Bleazard, T., Park, H.S., Rhee, H., et al.: FX: an RNA-Seq analysis tool on the Cloud. Bioinformatics **28**, 721–723 (2012)
35. Langmead, B., Hansen, K.D., Leek, J.T., et al.: Cloud-scale RNA-sequencing differential expression analysis with Myrna. Genome Biol. **11**, 1–11 (2010)
36. Zhang, L., Gu, S., Liu, Y., Wang, B., Azuaje, F.: Gene set analysis in the Cloud. Bioinformatics **28**, 294–295 (2012)

37. Almeida, J.S., Grüneberg, A., Maass, W., Vinga, S.: Fractal MapReduce decomposition of sequence alignment. Algorithms Mol. Biol. **7**, 1–12 (2012)
38. Cattaneo, G., Ferraro Petrillo, U., Giancarlo, R., Roscigno, G.: An effective extension of the applicability of alignment-free biological sequence comparison algorithms with Hadoop. J. Supercomput. 1–17 (2016). http://dx.doi.org/10.1007/s11227-016-1835-3
39. Hill, C.M., Albach, C.H., Angel, S.G., Pop, M.: K-mulus: strategies for BLAST in the Cloud. In: Wyrzykowski, R., Dongarra, J., Karczewski, K., Waśniewski, J. (eds.) PPAM 2013. LNCS, vol. 8385, pp. 237–246. Springer, Heidelberg (2014). doi:10.1007/978-3-642-55195-6_22
40. Colosimo, M.E., Peterson, M.W., Mardis, S., Hirschman, L.: Nephele: genotyping via complete composition vectors and MapReduce. Source Code Biol. Med. **6**, 1–10 (2011)
41. Drew, J., Hahsler, M.: Strand: fast sequence comparison using MapReduce and locality sensitive hashing. In: Proceedings of the 5th ACM Conference on Bioinformatics, Computational Biology, and Health Informatics, pp. 506–513. ACM (2014)
42. Chang, Y.J., Chen, C.C., Chen, C.L., Ho, J.M.: A de novo next generation genomic sequence assembler based on string graph and MapReduce Cloud computing framework. BMC Genomics **13**, 1–17 (2012)
43. Schatz, M.C., Sommer, D., Kelley, D., Pop, M.: De novo assembly of large genomes using Cloud computing. In: Proceedings of the Cold Spring Harbor Biology of Genomes Conference (2010)
44. Schatz, M.C.: BlastReduce: high performance short read mapping with MapReduce. University of Maryland (2008). http://cgis.cs.umd.edu/Grad/scholarlypapers/papers/MichaelSchatz.pdf
45. Schatz, M.C.: CloudBurst: highly sensitive read mapping with MapReduce. Bioinformatics **25**, 1363–1369 (2009)
46. Pireddu, L., Leo, S., Zanetti, G.: SEAL: a distributed short read mapping and duplicate removal tool. Bioinformatics **27**, 2159–2160 (2011)
47. Zhao, G., Ling, C., Sun, D.: SparkSW: scalable distributed computing system for large-scale biological sequence alignment. In: 15th IEEE/ACM International Symposium on Cluster, Cloud and Grid Computing (CCGrid), pp. 845–852. IEEE (2015)
48. Leo, S., Santoni, F., Zanetti, G.: Biodoop: bioinformatics on Hadoop. In: International Conference on Parallel Processing Workshops (ICPPW 2009), pp. 415–422. IEEE (2009)
49. Radenski, A., Ehwerhemuepha, L.: Speeding-up codon analysis on the Cloud with local MapReduce aggregation. Inf. Sci. **263**, 175–185 (2014)
50. Rasheed, Z., Rangwala, H.: A Map-Reduce framework for clustering metagenomes. In: IEEE 27th International Parallel and Distributed Processing Symposium Workshops & Ph.D. Forum (IPDPSW), pp. 549–558. IEEE (2013)
51. Matthews, S.J., Williams, T.L.: MrsRF: an efficient MapReduce algorithm for analyzing large collections of evolutionary trees. BMC Bioinform. **11**, 1–9 (2010)
52. Feng, X., Grossman, R., Stein, L.: PeakRanger: a Cloud-enabled peak caller for ChIP-seq data. BMC Bioinform. **12**, 1–11 (2011)

Photogrammetric Meshes and 3D Points Cloud Reconstruction: A Genetic Algorithm Optimization Procedure

Vitoantonio Bevilacqua[1]([✉]), Gianpaolo Francesco Trotta[2], Antonio Brunetti[1],
Giuseppe Buonamassa[3], Martino Bruni[1], Giancarlo Delfine[1], Marco Riezzo[1],
Michele Amodio[1], Giuseppe Bellantuono[1], Domenico Magaletti[1],
Luca Verrino[1], and Andrea Guerriero[1]

[1] Department of Electrical and Information Engineering,
Polytechnic University of Bari, Via Orabona 4, 70125 Bari, Italy
vitoantonio.bevilacqua@poliba.it
[2] Department of Mechanical and Management Engineering,
Polytechnic University of Bari, Via Orabona 4, 70125 Bari, Italy
gianpaolofrancesco.trotta@poliba.it
[3] Apulia Makers 3D Srls, Via Giulio De Ruggiero 56, 70125 Bari, Italy
g.buonamassa@apuliamakers3d.it

Abstract. Virtual reconstruction of heritage is one of the most interesting and innovative tool for preservation and keeping of historical, architectural and artistic memory of many sites that are in danger of disappearing. Find the best way to present an object in virtual reality is necessary for reasons linked to technology itself. In particular, the rendering of heavy object, in terms of details and meshes, influences the presentation of the whole virtual scene. Different researches have shown the onset of problems such as sickness due to an incorrect construction and representation of virtual scenes. In this paper we propose a 3D points cloud reconstruction method based on an optimization procedure by using genetic algorithm to improve the mesh obtained by low cost acquisition devices. The improved photogrammetric technique could be used to build virtual scenario by inexpensive devices (i.e. smartphone), without the cost and computational complexity of expensive devices.

1 Introduction

Cultural heritage safekeeping and preservation is a topic of great relevance in all those countries with long history. Aging, extreme weather events and devaluation policies of the cultural heritage value by the local governments involve preservation and requalification actions. For the protection of cultural heritage, these actions should not reduce the reliability of cultural sites and they should not affect the local touristic industry. In the last decades, several projects and papers proposed ICT technologies for cultural heritage improvements and requalification. In particular Virtual Reconstruction and Augmented Reality (AR) results

© Springer International Publishing AG 2017
F. Rossi et al. (Eds.): WIVACE 2016, CCIS 708, pp. 65–76, 2017.
DOI: 10.1007/978-3-319-57711-1_6

the most interesting and innovative tools for preservation and keeping of historical, architectural and artistic memory of many sites that are in danger of disappearing [1]. These techniques can be used to improve the user experience on the site, overlapping the augmented content, such as information or 3D models, directly on the real objects. In this way, making users interactive on site, the fruition of a cultural content is more interesting and engaging [2]. Furthermore, thanks to Virtual Reality (VR), it is possible to show fictional or past environments, as well as monuments or cities, as they were in past ages. There are several techniques for the creation of a 3D model starting from real objects.

A first approach for the 3D reconstruction of this environment was attempted using photogrammetry [3]. This technique allows the definition of position, shape and dimensions of objects extracting information coming from photographic images appropriately captured through a low-cost procedure that generates a medium-high quality model in terms of precision and details. Another technique for 3D models reconstruction regards 3D scanners. 3D scanning of surfaces allows the capture of huge points cloud datasets that can be used in a Computer Aided Design environment to build accurate 3D models of meaningful objects in the reconstructed scene. This technique is generally expensive but generates very accurate model. The process is time consuming because, after a preliminary data cleaning and registration phase, a digital representation of the original surface has to be computed through a process of surface reconstruction that generates polygonal meshes. The choice between photogrammetry using low-cost device and 3D scanner leads to a trade-off between costs and accuracy. The goal of this work is to find a novel approach based on genetic algorithms for the quality improvement of 3D objects reconstructed by photogrammetric techniques.

2   Photogrammetry

There are several techniques for the 3D acquisition and reconstruction; in particular, they could be grouped into two categories: range-based [4] and image-based [5]. In particular, the most used range-based technique is laser scanning, while photogrammetry is the most used image-based technique.

2.1   Acquisition Techniques

A range-based technique uses an active optical sensor which is able to return a large number of 3D coordinates by measuring key distances on the object to reconstruct. This technique reaches micron resolutions but it needs complex algorithms for the reconstruction of the 3D points cloud [6]. According to literature [7], the work-flow of a real object acquisition using an active optical sensor is:

- acquisition of different clouds of points or range-map in order to detect the whole scene;
- registration and alignment of data in a single reference system;
- reduction of noise and data errors in overlapping areas;

- creation of a structured polygon model;
- registration of image data (textures) to the geometric model.

An image-based technique uses images for the reconstruction of 3D models. Photogrammetry is the technique that allows the reconstruction of 3D objects from photographs. Starting from homologous points detected in the images, this technique is used to evaluate key metrics about size, shape and position of an object or scene [8,9]. Compared to laser scanning, which requires a dedicated acquisition hardware, the pictures for photogrammetrry could be acquired with a smartphone camera which generally is less costly then an active optical sensor; this make photogrammetry a candidate technique for a low cost 3D reconstruction. In order to achieve an optimal image acquisition process, some tips are listed below:

- images should be acquired with a good quality camera sensor, although smartphones camera could provide good results;
- it is important to take pictures all around the object. Each portion of the object must appear in at least three photos and each photo must have an overlapping margin of about 60% with the near ones;
- it is recommended to make several close-up shots to photograph particular portions of the surface (e.g. decorations);
- zoom should not be used or at least it is necessary to maintain the same level of zoom for all the photos;
- it is better to take a lot of pictures to eventually select the best and to discard the others;
- it is necessary to avoid different light condition and different day time (in case of object exposed to sun light);
- photo resolution has an important impact on the computational complexity for the model generation (higher resolution requires higher computing power).

After images acquisition, the work-flow to evaluate key metrics of the acquired object or scene is [10]:

- camera calibration to evaluate internal orientation;
- triangulation of images to evaluate external orientation;
- creation of 3D scene to derive an unstructured, sparse or dense points cloud;
- creation of a structured polygon model.

The output of the steps described above is the mesh representing all the key points evaluated (see Sect. 2.2). Table 1 compares the image-based and range-based modelling features [10]. All the aspects shown in the Table 1 make photogrammetry an extremely low cost technique which returns a good model in terms of resolution and quality. Range-based modelling shows some limits on the 3D reconstruction of architectural structures, such as houses or churches; on the other hand, it is possible to use aerial photogrammetry, through an Unmanned Aerial Vehicle (UAV), that allows to obtain a set of images necessary to the next processing.

Table 1. Difference between image-based and range-based modelling features

Features	Photogrammetry (image-based modeling)	Laser scanner (range-based modelling)
Equipment costs	Contained	High
Manoeuvrability	Excellent	Sufficient
Acquisition time	Short	Long
Modelling time	Long	Long
3D informations	To evaluate	Evaluated
Distance dependency	Independent	Dependent
Dimension dependency	Independent	Dependent
Material dependency	Independent	Dependent
Light conditions dependency	Dependent	Independent only for Time-of-Flight systems [11]
Geometry dependency	Dependent	Independent
Texture dependency	Dependent	Independent
Scale	To provide	Implied (1:1)
Volume of generated data	Dependent on the resolution of the images and the type of measures	Dense cloud of points
Fine details modelling	Good	Excellent
Texture	Included	Low resolution
Metric survey of edges	Excellent	Not sufficient
Quantitative and statistics analysis	For each evaluated point	Total
Open source software	Few	Very few

2.2   Reconstruction Techniques

The main technique used to reconstruct a 3D object starting from photogrammetric images is the Structured-from-Motion (SfM) [12]. This method allows to automatically reconstruct a three-dimensional scene from a set of two-dimensional digital images. Figure 1 shows the whole process of 3D object reconstruction. The SfM technique is based on automatic detection of key points (features) in three or more images using the Scale-Invariant Feature Transform (SIFT) [13] or similar algorithm, which are used to perform an image matching. Then, a bundle adjustment procedure is performed [14] to evaluate both camera focal length (internal orientation) and the shot position for each image (external orientation). The output of this step is a set of key points coordinates used to reconstruct a sparse points cloud of the acquired object. A dense points cloud is then generated increasing the number of neighbours of each element of the sparse points cloud. Finally, a 3D model is obtained by converting a simple points cloud in a set of vertices and faces that correspond to a mesh. In this

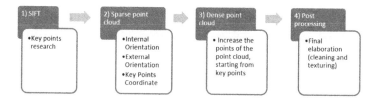

Fig. 1. 3D object reconstruction process using SfM method

Table 2. Qualitative comparison between obtained 3D model

Parameters [# points]	PPT + meshlab	Photoscan	Remake
Sparse points cloud	37045	3915	unknown
Dense points cloud	167228	1702888	unknown
Mesh	396760	168493	117313

final phase, the model may be also cleaned and textured. A qualitative comparison between the obtained models, based on the number of points in the sparse points cloud, dense points cloud and after the mesh registration, was performed (Table 2); to do this, three low-cost software have been compared:

- Python^{TM} Photogrammetric Toolbox (PPT): Free software with local elaboration
- Photoscan: Commercial software with local elaboration (used in 30 days free demo mode)
- Autodesk® Remake: Commercial software with remote elaboration (used in free education license)

The points cloud obtained from the three software are shown in the Fig. 2.

3   Mesh Improvement

After the comparison shown in the previous section, we chose the object reconstructed using Autodesk® Remake software because it was the best trade-off in terms of software cost and quality of rendering. Since photogrammetric systems are cheaper than range-based systems, in this work we try to compare the models obtained by using different photogrammetric sensors; in particular we used a smartphone camera and the Artec Eva 3D scanner. As shown in the images (Fig. 3(a) and (b)), the photogrammetric objects obtained with the two sensors show a great difference in terms of quality of rendering. As could be seen, the best quality is reached using the 3D scanner, but this technique is both time consuming and more expensive than a smartphone which is accessible to all; moreover, 3D scanner reconstruction needs a computer with higher computational capabilities. Our objective is to apply subdivision algorithms to improve the quality of the mesh acquired by photogrammetry using low-cost device so

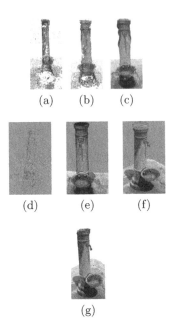

(a)   (b)   (c)

(d)   (e)   (f)

(g)

Fig. 2. (a) PPT Sparse Cloud (b) PPT Dense Cloud (c) PPT Mesh (d) PhotoScan Sparse Cloud (e) PhotoScan Dense Cloud (f) PhotoScan Mesh (g) Remake Mesh

(a)   (b)

Fig. 3. Comparison between the two techniques: (a) object reconstructed using photogrammetric technique with photos by smartphone camera; (b) object reconstructed using Artec Eva 3D Scanner

that the resulting 3D model has similar accuracy and resolution as a 3D-scanned model.

3.1   Subdivision Algorithms

The Surface Subdivision allows the surface smoothing by polygonal meshes. The final smooth surface is computed from the initial mesh by the recursive subdivision of each polygonal face into smaller faces that better approximate the smooth surface. The Surface Subdivision is based on Chaikin algorithm [15]: starting from an initial curve with certain points $P_0, P_0,...,P_n$, new vertices are created between points P_i, P_{i+1} for each subdivision step. The new points are computed with Eqs. 1 and 2:

$$q_{2i}^{k+1} = \frac{3}{4}p_i^k + \frac{1}{4}p_{i+1}^k \tag{1}$$

$$q_{2i+1}^{k+1} = \frac{1}{4}p_i^k + \frac{3}{4}p_{i+1}^k \tag{2}$$

Finally, the new curve is created using only the new vertices, as shown in Fig. 4

Limit curve   First step of subdivision

Fig. 4. Result of Chaikin algorithm

A subdivision algorithm can be classify according to some parameters:

- **Subdivision rules**
- Primal (Fig. 5(a)-(b)): a new vertex for every edge of the given mesh is inserted; then, the new vertices are connected [16];
- Dual (Fig. 5(c)-(d)): each vertex is divided to create new vertices for each near edge of the first vertex.
- **Subdivision types**
- Static: for each subdivision step the same rule is used;
- Dynamic: for each subdivision step different rules are used.
- **Mesh type**
- Triangular;
- Quadrangular.
- **Subdivision methods**
- Approximation method: if the final mesh does not contain initial vertices;
- Interpolation method: if the final mesh contains initial vertices.

These algorithms consider also the smoothness as the continuity property of limit surfaces (C^0, C^1, ..., C^n) [17]. In this work, the algorithms used to implement the surface subdivision are based on primal rule; they are reported in Table 3.

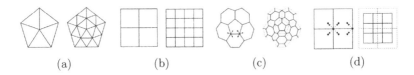

Subdivision Rules. (a) Primal schema for triangular mesh (b) Primal schema for quadrangular mesh (c) Dual schema for triangular mesh (d) Dual schema for quadrangular mesh

Table 3. Algorithm used to implement the surface subdivision

Algorithm	Methods	Mesh type	Mesh Continuity
Butterfly [18]	Interpolation	Triangular	C^1
Catmull-Clark [19]	Approximation	Triangular-Quadrangular	C^2
LS3Loop [20]	Approximation	Triangular-Quadrangular	C^2
Loop [21]	Approximation	Triangular	C^2
Midpoint [22]	Approximation	Triangular-Quadrangular	C^1

4   Genetic Algorithm

An optimization problem can be solved using one of different methods proposed in literature. Many of them are inspired from natural processes. This kind of methods, called evolutionary computation, generally starts with an initial population that evolve to achieve the global optimum of an objective function. The most popular evolutionary technique is Genetic Algorithm (GA) that uses operators that come from genetic variation and natural selection [23]. The firsts implementations of Genetic Algorithms can be found in late 1950's [24,25]. However, this field doesn't emerge since 1980's when the computational power of computers starts to increase and algorithms were improved [23]. To improve the mesh obtained by photogrammetric acquisition using a low-cost device in terms of resolution and accuracy, a multi-objective genetic algorithm [26,27] was designed. Our genetic algorithm is designed to find the best combination of subdivision algorithm, and its parameters, to apply to the model obtained by photogrammetry in order to reduce the distance between this model and a gold standard represented by the object obtained from 3D scanner. The workflow of the proposed approach is developed through the following steps:

1. initialization of the population of the GA;
2. the fitness function is evaluated for each element of the population:
   (a) application of a surface subdivision algorithm to the photogrammetric model;
   (b) Hausdorff distance evaluation between the original model and the model obtained in previous step;
3. application of selection and mutation to the population;
4. fitness function evaluation for the new individuals:

(a) application of a surface subdivision algorithm to the photogrammetric model;

(b) Hausdorff distance evaluation between the original model and the model obtained in previous step;

5. repeat from step 3 until the satisfaction of one of the output conditions;

6. Hausdorff distance evaluation between the model returned by 3D scanners and the photogrammetric model processed with the combination of surface subdivision algorithm and parameters resulting from GA.

Where:

- the population is randomly initialized;
- the selection operator is tournament with size equals to 3;
- the crossover technique is two-point crossover;
- the mutation can be performed on each bit of the gene.

The fitness function is composed by two objectives:

- number of points: points number of the cloud obtained after the execution of a subdivision surface algorithm;
- Hausdorff distance [28]: the distance between the mesh obtained after the subdivision surface algorithm elaboration and the initial mesh (see Sect. 4.1).

Each individual is described by a chromosome with three genes coding:

- X_1 - the surface subdivision algorithm (in Butterfly, Catmull-Clark, LS3Loop, Loop, Midpoint);
- X_2 - number of surface subdivision algorithm iterations (an integer ranging in $[0, 3]$);
- X_3 - percentage threshold (an integer ranging in $[0, 7]$) evaluated considering Eq. 3, which is the percentage of the diagonal of a bounding box containing the object mesh.

$$threshold = \frac{X_3 + 1}{8} \qquad (3)$$

Operators employed in the GA set up were:

- generations limit number: 100;
- crossover with a probability of 0.5;
- mutation with probability of 0.2 for each individual;
- each bit can be mutated with probability of 0.05;
- individuals for each generation: 50;
- stop criteria: maximum generations numbers (100) or 20 consecutive generations with the same fitness score.

4.1   The Hausdorff Distance

Considering two meshes and their vertices, for each vertex of the first land-mark mesh (X) the distance from the closest vertex in the second comparison mesh (Y) is evaluated; the output of this algorithm is a list of distances. In this work, we used the algorithm of Hausdorff distance implemented in Meshlab, an advanced 3D mesh processing software system that is oriented to the management and processing of unstructured large meshes and provides a set of tools for editing, cleaning, healing, inspecting, rendering, and converting these kinds of meshes. The output of the Meshlab implementation of Hausdorff distance is the maximum distance found in the comparison process (Eq. 4).

$$\sup_{x \in X} \inf_{y \in Y} d(x, y) \tag{4}$$

In general, Hausdorff distance is a symmetrical measure $(d(X, Y) = d(Y, X))$; in detail, its definition is in Eq. 5.

$$d_H(X, Y) = max\{\sup_{x \in X} \inf_{y \in Y} d(x, y), \sup_{y \in Y} \inf_{x \in X} d(x, y)\} \tag{5}$$

5   Results

The initial mesh, which is the input of the genetic algorithm, acquired by a smartphone camera and reconstructed by Autodesk® Remake Educational Edition software, is characterised by 17027 vertices (bounding box diagonal = 1390.367432). The mesh used as ground truth, acquired by the Artec Eva 3D Scanner and reconstructed by Artec Eva image-based modelling commercial software, is characterised by 51701 vertices (bounding box diagonal = 1403.514160).

The Hausdorff distance, used to compare the differences between the previous two meshes, considering 3D scanner mesh as reference, has a mean value of 3.778587 (min: 0.000366, Max: 34.477371, RMS: 5.043168). The designed multi-objective genetic algorithm output is reported in Table 4.

Table 4. The best individual from the GA

Surface Subdivision Algorithm	Vertices Number	Iterations Number	Percentage Threshold	Hausdorff distance with percentage variation			
Butterfly	75137 (+441.281%)	3	0.5	Min	Max	Mean	RMS
				0.000092 (−74.863%)	34.534580 (+0.166%)	3.778014 (−0.015%)	5.043389 (+0.004%)

The subdivision algorithm is not computationally intensive because its execution is 473 ms, on a system configured with a CPU Intel® Core™i3-2350M (2.3 GHz, 3 MB L3 cache), an integrated GPU Intel® HD Graphics 3000 and

4 GB DDR3 Memory. The percentage variation in the results table represents the Hausdorff distance variation between the difference obtained comparing 3D-scanned and the initial photogrammetric meshes and comparing 3D-scanned and the genetic algorithm output meshes.

6 Discussion and Conclusion

The results obtained show that the surface subdivision algorithms increases the points number of mesh (+441.281%), while the relative Hausdorff distance values show a mesh improvement as report the percentage reduction of the distances mean (−0.015%). Considering these results, there was a little mesh improvement of 3D object quality but it was not considerable in terms of object rendering. Probably it depends on the position of the points generated by the surface subdivision algorithms which are randomly placed by the algorithms themselves. To overcome this issue, in future work we will try to better design the GA taking into account new parameters, such as the points positioning and distribution. Another possible way to improve the performance of our genetic algorithm is to apply a different method for maintaining the population: the steady state approach [23]; in fact, this allows the replacement of worst, or oldest, individuals in the population by children as soon as they are created.

References

1. Guttentag, D.A.: Virtual reality: applications and implications for tourism. Tour. Manag. **31**(5), 637–651 (2010)
2. Fritz, F., Susperregui, A., Linaza, M.T.: Enhancing cultural tourism experiences with augmented reality technologies. In: 6th International Symposium on Virtual Reality, Archaeology and Cultural Heritage (VAST) (2005)
3. Suveg, I., Vosselman, G.: 3D reconstruction of building models. Int. Arch. Photogrammetry Remote Sens. **33**(B2; PART 2), 538–545 (2000)
4. Tangelder, J.W.H., Veltkamp, R.C.: A survey of content based 3D shape retrieval methods. Multimedia Tools Appl. **39**(3), 441–471 (2008)
5. Remondino, F., El-Hakim, S.: Image-based 3D modelling: a review. Photogram. Rec. **21**(115), 269–291 (2006)
6. Bevilacqua, V., Ivona, F., Cafarchia, D., Marino, F.: An evolutionary optimization method for parameter search in 3D points cloud reconstruction. In: Huang, D.-S., Bevilacqua, V., Figueroa, J.C., Premaratne, P. (eds.) ICIC 2013. LNCS, vol. 7995, pp. 601–611. Springer, Heidelberg (2013). doi:10.1007/978-3-642-39479-9_70
7. Guidi, G., Russo, M., Beraldin, J.A.: Acquisizione 3D e modellazione poligonale. McGraw-Hill, New York (2009)
8. Mikhail, E.M., Bethel, J.S., McGlone, J.C.: Introduction to modern photogrammetry, New York (2001)
9. Luhmann, T., Robson, S., Kyle, S., Harley, I.: Close range photogrammetry: principles, methods and applications. Whittles (2006)
10. Russo, M., Remondino, F.: Laser scanning e fotogrammetria: strumenti e metodi di rilievo tridimensionale per larcheologia. APSAT 1, 133–164 (2012)
11. Hagebeuker, B.: A 3D time of flight camera for object detection (2007)

12. Westoby, M.J., Brasington, J., Glasser, N.F., Hambrey, M.J., Reynolds, J.M.: structure-from-motion photogrammetry: a low-cost, effective tool for geoscience applications. Geomorphology **179**, 300–314 (2012)
13. Lowe, D.G.: Object recognition from local scale-invariant features. In: The proceedings of the seventh IEEE international conference on Computer vision, vol. 2, pp. 1150–1157. IEEE (1999)
14. Triggs, B., McLauchlan, P.F., Hartley, R.I., Fitzgibbon, A.W.: Bundle adjustment — a modern synthesis. In: Triggs, B., Zisserman, A., Szeliski, R. (eds.) IWVA 1999. LNCS, vol. 1883, pp. 298–372. Springer, Heidelberg (2000). doi:10. 1007/3-540-44480-7_21
15. Chaikin, G.M.: An algorithm for high-speed curve generation. Comput. Graph. Image Process. **3**(4), 346–349 (1974)
16. Kobbelt, L.: 3-subdivision. In: Proceedings of the 27th Annual Conference on Computer Graphics and Interactive Techniques, pp. 103–112. ACM Press/Addison-Wesley Publishing Co. (2000)
17. Micchelli, C.A., Prautzsch, H.: Uniform refinement of curves. Linear Algebra Appl. **114**, 841–870 (1989)
18. Dyn, N., Levine, D., Gregory, J.A.: A butterfly subdivision scheme for surface interpolation with tension control. ACM Trans. Graph. **9**(2), 160–169 (1990)
19. Catmull, E., Clark, J.: Recursively generated b-spline surfaces on arbitrary topological meshes. Comput. Aided Des. **10**(6), 350–355 (1978)
20. Guennebaud, S.B., Schlick, C.: Least squares subdivision surfaces. Comput. Graph. Forum **29**(7), 2021–2028 (2010)
21. Loop, C.: Smooth subdivision surfaces based on triangles (1987)
22. Habib, A., Warren, J.D.: Edge and vertex insertion for a class of C1 subdivision surfaces. Comput. Aided Geom. Design **16**(4), 223–247 (1999)
23. Sivanandam, S.N., Deepa, S.N.: Introduction to Genetic Algorithms. Springer Science & Business Media, Heidelberg (2008)
24. Friedberg, R.M.: A learning machine: Part I. IBM J. Res. Dev. **2**(1), 2–13 (1958)
25. Friedberg, R.M., Dunham, B., North, J.H.: A learning machine: Part II. IBM J. Res. Dev. **3**(3), 282–287 (1959)
26. Bevilacqua, V., Brunetti, A., Triggiani, M., Magaletti, D., Telegrafo, M., Moschetta, M.: An optimized feed-forward artificial neural network topology to support radiologists in breast lesions classification. In: Genetic and Evolutionary Computation Conference, GECCO 2016, pp. 1385–1392. ACM (2016)
27. Bevilacqua, V., Mastronardi, G., Menolascina, F., Pannarale, P., Pedone, A.: A novel multi-objective genetic algorithm approach to artificial neural network topology optimisation: the breast cancer classification problem. In: Proceedings of the International Joint Conference on Neural Networks, IJCNN 2006, pp. 1958–1965. IEEE (2006)
28. Aspert, N., Santa-Cruz, D., Ebrahimi, T.: MESH: measuring errors between surfaces using the hausdorff distance. In: Proceedings of the 2002 IEEE International Conference on Multimedia and Expo, vol. I, pp. 705–708. IEEE Computer Society (2002)

Benchmarking Spark Distributed Data Structures: A Sequence Analysis Case Study

Umberto Ferraro Petrillo[(✉)] and Roberto Vitali

Università di Roma "La Sapienza", 00185 Roma, Italy
{umberto.ferraro,roberto.vitali}@uniroma1.it

Abstract. Big Data technologies are recognized by many as a promising solution for the efficient management and analysis of the enormous amount of genomic data available thanks to Next-Generation Sequencing technologies. Despite this, they are still used in a limited number of cases, mostly because of their complexity and of their relevant hidden computational constants. The introduction of Spark is changing this scenario, by delivering a framework that can be used to write very complex and efficient distributed applications using only few lines of codes. Spark offers three types of distributed data structures that are almost functionally equivalent but are very different in their implementations. In this paper, we briefly review these data structures and analyze their advantages and disadvantages, when used to solve a paradigmatic bioinformatics problem on a Hadoop cluster: the k-mer counting.

Keywords: Spark · k-mers Counting · Distributed computing · Performance analysis

1 Introduction

The introduction of next-generation sequencing technologies (in short, NGS) has changed the landscape of biology [1,2], thanks to the possibility of sequencing DNA at a much faster speed than the one achievable with traditional Sanger sequencing approach [3]. This advancement has raised also new methodological and technological challenges. One of these is about the proper approach to adopt for managing and processing timely the vast amount of data that is produced thanks to NGS technologies.

A solution that is gaining popularity is to resort to the technologies that have been developed for dealing with Big Data. By this term, we refer to the problem of storing, managing and processing data that may be significantly big with respect to several dimensions like size, diversity or generation rate. A very popular approach to Big Data processing, allowing for the analysis of enormous datasets, is the one based on MapReduce [4]. It is a computational paradigm that works by organizing an elaboration in two consecutive steps. In the first step, a *map* function is used to process, filter and/or transform input data. In the second step, a *reduce* function is used to aggregate the output of the map functions. Map

© Springer International Publishing AG 2017
F. Rossi et al. (Eds.): WIVACE 2016, CCIS 708, pp. 77–88, 2017.
DOI: 10.1007/978-3-319-57711-1_7

and reduce functions are executed as tasks on the nodes of a distributed system, namely, a network of computational nodes that cooperate, sending messages each other, to achieve a common goal. The most used implementation of this paradigm is Apache Hadoop [5]. Despite being the first framework to provide a full implementation of the MapReduce paradigm, Hadoop is often criticized for a number of issues, first being its disappointing performance when used for running iterative tasks (see, e.g., [6]). A competing framework is gaining a lot of attention in the very recent years: Apache Spark [7]. It is a sort of evolution of Hadoop, but with some important differences allowing it to outperform its predecessor in many application scenarios. First of all, wherever there is enough RAM, Spark is able to perform iterative computations in-memory, without having to write intermediate data on disk, as required by Hadoop. In addition, it is more flexible than Hadoop, because it provides a rich set of distributed operations other than the ones required for implementing the MapReduce paradigm.

Indeed, one of the aspects that has the deepest impact on the performance of a distributed application, is the pattern used to distribute and process data among the different nodes of a network. This is especially the case of bioinformatics application, where even a single genomic sequence may be several gigabytes long. In this context, a poor data layout may prevent even an efficient algorithm to exploit the parallelism of a distributed system. From this viewpoint, the Spark feature that most marks the difference with respect to Hadoop is the availability of ready-to-use distributed data structures. These allow to manage and process in a standard and consistent way the data of an application while leaving to Spark the responsibility of partitioning this data and their elaboration. It is interesting to note that Spark offers three different types of distributed data structures. These are almost functionally identical and choosing which of them to use when developing a bioinformatics application may not be simple.

The goal of this paper is to investigate the complexity and the performance of the different distributed data structures offered by Spark, with the aim of providing useful hints to the bioinformatics community about which is the best option to choose, and when. This has been done by analyzing the three different solutions when used for the implementation of a typical sequence analysis algorithmic pattern: the counting of the distinct k-mers existing in an input sequence of characters. The three implementations we developed have been tested on a reference dataset to determine their relative performance and provide insightful hints about which of them to prefer when dealing with such a scenario.

Organization of the paper. The paper is structured as follows. In Sect. 2, we briefly discuss the current state of adoption of Big Data technologies for genomic computation. The Spark framework and the distributed data structures it offers are presented in Sect. 3. In Sect. 4, we state the objective of this paper and present the k-mer counting problem that has adopted as reference scenario for evaluating the different types of Spark distributed data structures. In Sect. 5 we outline the setting of our experiments and briefly discuss their results. Finally, in Sect. 6 we provide some concluding remarks for our work.

2   Related Work

The adoption of Big Data related technologies for accelerating the solution of bioinformatics problems has proceeded at a slow pace in the past years for several reasons. One of these is that the complexity of a framework like Hadoop adds to a distributed computation a significant amount of overhead, thus making it convenient only when processing enormous amount of data and/or when using distributed facilities counting hundreds or thousands of nodes. Instead, the same computation carried on a stand-alone workstation is able to exploit almost all the processing power of the underlying machine as the logic required to coordinate several concurrent processes running on the same machine is much simpler. Despite this, there are several relevant contributions worth to be mentioned.

One of the first and most noteworthy is GATK (see [8]). It introduces a structured programming framework designed to ease the development of efficient and robust analysis tools for next-generation DNA sequencers using MapReduce. A problem that often arises when writing an Hadoop sequence analysis application is the adaptation of the formats used for maintaining genomic sequences to the standard file format used by Hadoop. This problem has been addressed by Niemenmaa et al. in [9]. They proposed a software library for the generic and scalable manipulation of aligned next-generation sequencing data stored using the BAM format. The same problem has been further addressed by Massie et al. in [10]. In this case, the authors did not resort to an existing file format, like in [9], but proposed a new file format (i.e., ADAM) explicitly designed for indexing and managing genomic sequences on a distributed MapReduce system like Hadoop or Spark. There have also been several contributions about the usage of MapReduce and Hadoop for the solution of specific application problems. To name some, the work by Cattaneo et al. in [11,12] describes a MapReduce distributed framework based on Hadoop able to evaluate the alignment-free distance among genomic sequences using a variety of dissimilarity functions and in a scalable way.

The advent of Spark is slowly changing this scenario, as there is an increasing number of contributions developed using this technology and aiming at introducing solutions that are not only scalable but also efficient. This is the case of SparkSeq, a general-purpose, flexible and easily extendable library for genomic cloud computing presented in [13]. It can be used to build genomic analysis pipelines and run them in an interactive way. Another work worth to be mentioned is the one described in [14]. It provides a comprehensive study on a set of distributed algorithms implemented in Spark for genomic computation adopting efficient statistical approaches. The main objective is the study of the performance of the proposed algorithms with respect to more traditional ones.

3   Spark

Spark is a framework for general-purpose distributed processing of Big Data. It is able to exploit the computational capabilities of several calculators at once, by providing an uniform and abstract view of these as a computing cluster. Spark

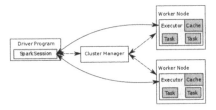

Fig. 1. Spark architecture

can be seen as a sort of evolution of Hadoop, as inherits the same MapReduce based distributed programming paradigm. In addition, it offers a wide range of ready-to-use operators and transformations that are often needed when developing a distributed application and that, although being possible with Hadoop, would require some work to be developed from scratch.

The Spark architecture (see Fig. 1) is composed by three main components: (a) the *driver program*, that is in charge to setup the Spark environment and launch the computation; (b) a *cluster manager* service, that is in charge of managing the distributed computation, assigning resources and scheduling the execution of one or more *tasks* on each node of the cluster; (c) several different *worker nodes*, in charge of carrying out the real computation, where each node is able to execute one or more tasks in parallel by spanning a corresponding number of *executors*. Notice that, apart from the cluster manager available with Spark, it is also possible to use third-party managers, such as Hadoop *YARN* [5].

3.1 The Programming Model

In a typical Spark application, the *driver program* begins the execution by loading the input data in a distributed data structure. This is essentially a collection of objects that is partitioned over the nodes of a cluster. Once data has been loaded, the execution proceeds by means of a sequence of distributed operations. Following the same *move computation close to data* philosophy that inspired Hadoop, Spark tries to run these operations directly on the nodes hosting the data that they are required to process. This is done to reduce the overhead that will be otherwise required to transfer big amount of data over the network for processing them elsewhere.

Distributed data structures available with Spark support two types of distributed operations: *Actions* and *Transformations*. The former may essentially be divided in three categories:

- **reduce:** apply a cumulative operation to the elements of a distributed collection of objects, so that multiple input objects are aggregated and combined in a single object belonging to an output distributed collection of objects;
- **collect:** gather all the objects of a distributed collection, or a subset of them, and send them to the driver program, where these are made available as a collection of local objects;

- **save:** writes the elements of distributed collection of objects on an external storage.

The latter may essentially divided in the three categories:

- **map:** map a distributed collection of objects in another distributed collection of objects. The new objects result from the application of a given function on each of the input objects;
- **filter:** filter the elements of a distributed collection of objects, returning a new distributed collection containing only elements satisfying an input-provided condition;
- **set operations:** combine two distributed collection of objects in a single one by means of a set operator.

The distributed part of an application run with Spark is logically divided in *stages*, where each stage corresponds to a transformation or an action. Stages related to transformations are run by Spark in a *lazy* way. This means that they are not run as soon as they are encountered during the execution of a program but only when and if their result is needed to accomplish a subsequent step of the application.

3.2   Distributed Data Structures

Spark provides three types of distributed data structures: *Resilient Distributed Dataset*, *DataFrame* and *DataSet*. These data structures share some relevant properties. First, they do all support in-memory computations. This means that, provided that there is enough memory space, their content may be partially or entirely cached in memory. This is especially useful when executing subsequent or iterative tasks targeting the same data. If the available memory is not enough, as when processing very large amount of data, their content may be selectively spilled to disk and retrieved in memory when required. The developer can choose also if and how to replicate this data, so to make the computation resilient with respect to hardware or network faults (see [7] for examples).

Resilient Distributed Dataset. The Resilient Distributed Dataset (in short, RDD) has been the first type of distributed data structure available with Spark. It is a virtual data structure encapsulating a collection of object-oriented datasets spread over the nodes of a computing cluster. The object-oriented nature of these datasets implies all the advantages and the disadvantages of this paradigm. For instance, it is the developer that chooses how the data stored in a RDD can be processed, by defining some proper methods on the objects storing that data. RDDs can be created by importing a dataset from an external storage or from the network, by issuing some special-purpose functions provided by Spark for making distributed a local dataset or as the result of the execution of a transformation over another RDD.

RDDs have also some drawbacks. For example, every time there is need to transfer elsewhere the content of a RDD (e.g. when performing a reduce operation), Spark has to marshall and encode, one-by-one, all the elements of that

RDD as well as their associated metadata. The reverse of this operation, then, has to be performed on each node receiving those elements. Similarly, whenever the content of a RDD is destroyed, the underlying java virtual machine has to claim back the memory used by each of the objects contained in the RDD. Since RDDs are often used to maintain collections counting millions or billions of elements, this overhead may severely burden the performance of a Spark application.

DataFrame. The `DataFrame` is a distributed data structure introduced in Spark to overcome some of the performance issues of RDDs. Instead of using a collection of objects, `DataFrames` maintain data in a relational-database fashion, providing a flat table-like representation supported by the definition of a schema. This has several important advantages. First, manipulation of large amount of data can be carried out using an SQL-like engine rather than requiring the execution of methods on each of the element to be processed. Second, by avoiding the usage of objects for storing the individual elements of a collection, the transmission of a chunk of a `DataFrame` to a node tends to be very fast. Third, since the metadata describing the elements of a collection are the same for all these elements and are known in advance, there is no need of transmitting them when moving parts of a `DataFrame`, thus achieving a substantial saving in communication time. Finally, the adoption of an SQL based approach to the processing of data allows for several optimizations (see [15]). Even `DataFrames` suffer of some serious drawbacks. To name one, the dismissal of the object-oriented approach in favor of the SQL-like engine makes the resulting applications less robust as it is virtually impossible for the compiler to verify the type-safety of an application.

DataSet. The `DataSet` is a distributed data structure introduced to mix the best of the two previous technologies by guaranteeing the same performance of `DataFrames` while allowing to model data after the object oriented paradigm, as when using RDDs. This is mainly achieved thanks to two solutions. The first is the introduction of a new *encoding* technology able to marshall quickly and in a step a collection of objects. We recall that RDD need to marshall individually each object of a collection by means of the Java standard serialization framework. The second is the possibility of operating on the elements of a `DataSet` using an object-oriented interface while retaining their internal relational representation. On a side, this allows to perform the safety checks at compile time, thus making the applications more robust. On the other side, this allow to maintain all the performance advantages introduced with `DataFrames`.

4   Objective of the Paper

The three types of distributed data structures available with Spark, as well as the wide range of transformations and actions they provide, allow to write complex distributed applications in a few lines of code and without requiring advanced programming skills. This is a relevant feature as, typically, one of major issues preventing from using a distributed approach to solve a problem is the time and the cost required to develop such a solution. However, this simplicity comes at a cost. By delegating to Spark most of the work about how to organize and process distributed data structures, the developer takes the risk of sacrificing the efficiency of his code.

In this paper, we deal with this problem by focusing on assessing the performance trade-offs related to the choice of the distributed data structure type among the three offered by Spark, when developing a bioinformatics application. We use as a case study a simple problem that is fundamental when performing genomic sequence analysis: the k-mer counting problem.

4.1 The k-mer Counting Problem

Given a string S, we denote with term *k-mer* all the possible substrings of S having size k. The k-mer counting problem refers to the problem of counting the number of occurrences of each *k-mer* k in S. It is a very common and (apparently) simple task that is often used as a building block in the development of more complex sequence analysis applications such as genome assembly or sequence alignment (see, e.g., [16]).

The problem of counting the k-mers of a sequence is paradigmatic with respect to the class of problems that would benefit from the adoption of a distributed solution. On a side, it is apparently easy to solve as its algorithmic formulation is very simple and straightforward. This simplifies as well its distributed reformulation, as it represents a typical case of an embarrassingly parallel problem. On the other side, real-world scenarios often require to process either a huge number of sequences or sequences having a huge size (i.e., gigabytes of characters). Consequently, there is both a time-related problem (i.e., processing huge amount of data using a single machine could require days or weeks) and a memory-related problem (i.e., the memory required to keep the k-mers counts may span also tens or hundreds of gigabytes when using large values of k and huge sequences). The convenience of this approach is also witnessed by the several scientific contributions proposed so far (see, e.g., [11,12,17,18]), introducing clever solutions for counting k-mers in a parallel or distributed setting.

5   Experimental Study

In our experimental study we first developed three different solutions to the k-mer counting problem using Spark. These solutions are identical in their output, provided the same input, but differ in the distributed data structures they use. Then, we performed a comparative experimental analysis of these codes by measuring their performance when run on a reference testing dataset.

5.1   The Proposed Implementations

We report in Listings 1.1, 1.2 and 1.3 the pseudo-code of our three implementations (full source code not reported and available upon request): RDD, DataSet

Listing 1.1. Pseudo-code of k-mer counting implemented using Resilient Distributed Datasets

```
1    input = readTextFile(filename);
2    kmers = input.flatMapToPair(new KMerExtractor());
3    kmers_count = kmers.reduceByKey(new KMerAggregator()
         );
4    writeFile(kmers_count);
```

and `DataFrame`. As already said, the three solutions are equivalent, except for the particular type of distributed data structure used by each of them.

The first solution (Listing 1.1) uses a RDD to collect all the string lines of an input file, where each line corresponds a different genomic sequence. Then, it applies to each line a map function, `KMerExtractor`, that scans it returning all the k-mers it contains as a RDD of pairs $(k\text{-}mer, 1)$ (line 2). All these pairs are aggregated by the `KMerAggregator` reduce function (line 3), thus returning a RDD containing the final counts. The result is saved to file (line 4).

The second solution (Listing 1.2) extracts the k-mers from an input file as the first solution (line 1–2). Then, it builds a new schema definition, needed to establish the structure of the `DataFrame` used for storing the k-mers (line 3). Then, a new `DataFrame` is created using this definition and the collection of extracted k-mers (line 4). Once the `DataFrame` is ready, it is queried through an SQL query (line 5–6) for the k-mer counts. The result is saved to file (line 7).

The third solution (Listing 1.3) mimics the second one, but without the need of defining an explicit schema. In details, it first extracts k-mers from an input file as in the previous cases (line 1–2). The results of the extraction is saved in a Dataset. Its schema is automatically determined according to the data type of the k-mers. Then, it is queried (line 3) by running some of the standard methods available with this data structure (i.e., `groupBy` and `count`), instead of using an SQL query. The k-mer counts resulting from the query is saved to file (line 4).

Listing 1.2. Pseudo-code of k-mer counting implemented using DataFrames

```
1    input = readTextFile(filename);
2    kmers = input.flatMap(new KMerExtractor());
3    schema = CreateNewSchema(schema definition);
4    createDataFrame("kmers", schema, kmers);
5    String q = "select kmer, count(kmer) as count from
         kmers group by kmer";
6    kmers_count = spark.sql(q);
7    writeFile(kmers_count);
```

Listing 1.3. Pseudo-code of k-mer counting implemented using DataSets

```
1    input = readTextFileasDataset(filename);
2    kmers = input.flatMap(new KMerExtractor());
3    kmers_count = kmers.groupBy("kmer").count();
4    writeFile(kmers_count);
```

5.2 Dataset

The experiments have been conducted on a dataset of four randomly-generated FASTA [19] files of increasing size. Each file has been generated as a collection of short-sequences, with each sequence being introduced by a text comment line and containing at most 100 characters drawn from the alphabet $\{A, C, G, T\}$. The overall size of the used files is, respectively, of about: 512 MB, 2 GB, 8 GB. These sizes have been chosen to represent the class of problems that are difficult to manage with a sequential approach and would benefit of a distributed solution.

5.3 Configuration

Our experiments have been conducted on a five-nodes Hadoop cluster, with one node acting as *resource manager* for the cluster and the remaining nodes being used as worker nodes. Each node of this cluster is equipped with a 16-core Intel Xeon E5-2630@2.40 GHz processor, with 64 GB of RAM. During the experiments, we varied the number of executors on each node from 1 to 4, to assess the scalability of the proposed solutions. Moreover, we organized input files in blocks having size at most 64MB, with each block available on two different nodes of the cluster. Such configuration allows for a better distribution of the workload but without affecting the performance of the whole system.

5.4 Results

In our first experiment, we have measured the performance of RDD, DataFrame and DataSet when run on sequences of increasing size and using increasing values of k. Its purpose has been to analyze the behavior of the three types of distributed data structures in a context where the size of these data structures could exceed the RAM memory available to a node. The experimental result, reported in Fig. 2, shows that when dealing with very small sized problems (i.e., size = 512 MB, k = 7) RDD is the implementation achieving the best performance. We recall that in this setting the number of possible distinct k-mers is very small. Consequently, RDD has to manage a very small number of objects. As soon as the size of the problem increases, the performance of this implementation quickly deteriorates because of the too many k-mers to be handled. Instead, the other two implementations exhibit an increase in their execution time that is linear with respect to the size of the problem. This is clearly due to their different strategy used to maintain k-mers in memory, that reveals to be much more

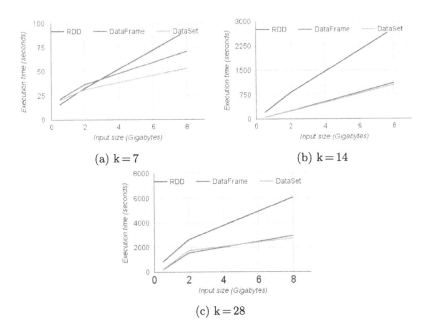

Fig. 2. Execution time, in seconds, of RDD, DataFrame and DataSet when processing random sequences of increasing size under different assignments of k

efficient when the number of k-mers to manage increases. We notice also that DataSet performs slightly better than DataFrame, mostly because it is able to encode k-mers faster (see Sect. 3.2).

In our second experiment, we have measured the scalability of the three considered implementations when run on a small problem instance and on a large problem instance using a cluster of increasing size. The two cases are representative of a scenario where the distributed data structures are either small enough to fit in the main memory or large enough to require their partial backup on external memory. The increasing size of the cluster has been simulated by increasing the number of executors per node (see Sect. 3), for an overall number of 4, 8 and 16 executors.

The experimental results on the small problem instance dataset, reported in Fig. 3, confirm that DataSet is the fastest of the three implementations. However, we notice that the scalability of RDD is much better. As expected, this phenomenon is due to the fact that, for such a small dataset, the usage of a high number of executors allows RDD to keep all the k-mers counts in memory, thus becoming competitive with the other two implementations. For the same reason, RDD enjoys a linear speed-up proportional to the number of executors. Instead, the performance of DataSet offers small room for improvement, as there is no noticeable gain when switching from 8 executors to 16 executors. Speaking of the large problem instance, we observe that none of the three codes is able to scale linearly with the number of executors. This may be explained by considering the

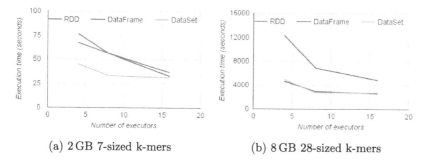

(a) 2 GB 7-sized k-mers   (b) 8 GB 28-sized k-mers

Fig. 3. Scalability of RDD, DataFrame and DataSet on a cluster with an increasing number of executors, when extracting k-mers from: (a) a 2 GB random sequences with k = 7; (b) a 8 GB random sequences with k = 28;

I/O bound nature of the k-mer counting activity, that becomes more evident when processing very large files. In such a scenario, most of the time is spent reading data from the external memory. Running several executors on the same node implies that they will contend the access to the disk when trying to read at the same time their respective input blocks, thus preventing the possibility of fully exploiting their computational resources.

6   Conclusion

The objective of this work has been to assess how the choice of the particular type of distributed data structure to be used for implementing a sequence analysis application with Spark affects its performance. We observed that three variants of the same code (a k-mer counting algorithm), having an identical behavior and undistinguishable in their output, but using different types of distributed data structures, exhibit very different performance. A direction worth to be investigated would be the analysis of more complex sequence analysis application patterns. This would allow to better assess the architectural peculiarities of the different types of the Spark distributed data structures. Moreover, given the internal complexity of Spark and the availability of a large number of settings influencing its performance, another promising direction would be to repeat these experiments on a larger scale and under a much broader range of configurations.

References

1. Pop, M., Salzberg, S.L.: Bioinformatics challenges of new sequencing technology. Trends Genet. **24**(3), 142–149 (2008)
2. Schuster, S.C.: Next-generation sequencing transforms today's biology. Nature **200**(8), 16–18 (2007)
3. Sanger, F., Nicklen, S., Coulson, A.R.: DNA sequencing with chain-terminating inhibitors. Proc. Natl. Acad. Sci. **74**(12), 5463–5467 (1977)

4. Dean, J., Ghemawat, S.: MapReduce: simplified data processing on large clusters. Commun. ACM **51**(1), 107–113 (2008)
5. Apache: Hadoop. http://hadoop.apache.org/
6. Zhang, Y., Gao, Q., Gao, L., Wang, C.: iMapReduce: a distributed computing framework for iterative computation. J. Grid Comput. **10**(1), 47–68 (2012). http://dx.doi.org/10.1007/s10723-012-9204-9
7. Apache: Spark. http://spark.apache.org/
8. McKenna, A., Hanna, M., Banks, E., Sivachenko, A., Cibulskis, K., Kernytsky, A., Kiran, G., Altshuler, D., Gabriel, S., Daly, M., DePristo, M.A.: The genome analysis toolkit: a MapReduce framework for analyzing next-generation DNA sequencing data. Genome Res. **20**(9), 1297–1303 (2010). http://genome.cshlp.org/content/20/9/1297.abstract
9. Niemenmaa, M., Kallio, A., Schumacher, A., Klemelä, P., Korpelainen, E., Heljanko, K.: Hadoop-BAM: directly manipulating next generation sequencing data in the cloud. Bioinformatics **28**(6), 876–877 (2012)
10. Massie, M., Nothaft, F., Hartl, C., Kozanitis, C., Schumacher, A., Joseph, A.D., Patterson, D.A.: ADAM: Genomics formats and processing patterns for cloud scale computing. University of California, Berkeley Technical report, No. UCB/EECS-2013 207 (2013)
11. Cattaneo, G., Ferraro-Petrillo, U., Giancarlo, R., Roscigno, G.: Alignment-free sequence comparison over Hadoop for computational biology. In: Proceedings of 44th International Conference on Parallel Processing Workshops, ICPPW, pp. 184–192 (2015)
12. Cattaneo, G., Ferraro-Petrillo, U., Giancarlo, R., Roscigno, G.: An effective extension of the applicability of alignment-free biological sequence comparison algorithms with Hadoop. J. Supercomputing. **73**(4), 1467–1483 (2017)
13. Wiewiórka, M.S., Messina, A., Pacholewska, A., Maffioletti, S., Gawrysiak, P., Okoniewski, M.J.: SparkSeq: fast, scalable, cloud-ready tool for the interactive genomic data analysis with nucleotide precision. Bioinformatics (2014)
14. Bahmani, A., Sibley, A.B., Parsian, M., Owzar, K., Mueller, F.: SparkScore: leveraging apache spark for distributed genomic inference. In: 2016 IEEE International Parallel and Distributed Processing Symposium Workshops, pp. 435–442, May 2016
15. Xin R., R.J.: Project tungsten: Bringing Spark closer to bare metal. https://databricks.com/blog/2015/04/28/project-tungsten-bringing-spark-closer-to-baremetal.html
16. Giancarlo, R., Rombo, S.E., Utro, F.: Epigenomic k-mer dictionaries: shedding light on how sequence composition influences in vivo nucleosome positioning. Bioinformatics (2015)
17. Deorowicz, S., Kokot, M., Grabowski, S., Debudaj-Grabysz, A.: KMC2: fast and resource-frugal k-mer counting. Bioinformatics **31**, 1569–1576 (2015)
18. Ferraro Petrillo, U., Roscigno, G., Cattaneo, G., Giancarlo, R.: FASTdoop: a versatile and efficient library for the input of FASTA and FASTQ files for MapReduce Hadoop bioinformatics applications. Bioinformatics (2017). https://dx.doi.org/10.1093/bioinformatics/btx010
19. Wikipedia: FASTA format – Wikipedia, the free encyclopedia. https://en.wikipedia.org/wiki/FASTA_format

Modelling and Simulation of Artificial and Biological Systems

Automatic Design of Boolean Networks for Cell Differentiation

Michele Braccini[1], Andrea Roli[1(✉)], Marco Villani[2,3], and Roberto Serra[2,3]

[1] Department of Computer Science and Engineering,
Alma Mater Studiorum Università di Bologna, Cesena, Italy
`andrea.roli@unibo.it`
[2] Department of Physics, Informatics and Mathematics,
Università di Modena e Reggio Emilia, Modena, Italy
[3] European Centre for Living Technology, Venice, Italy

Abstract. Cell differentiation is the process that denotes a cell type change, typically from a less specialised type to a more specialised one. Recently, a cell differentiation model based on Boolean networks subject to noise has been proposed. This model reproduces the main abstract properties of cell differentiation, such as the attainment of different degrees of differentiation, deterministic and stochastic differentiation, reversibility, induced pluripotency and cell type change. The generic abstract properties of the model have been already shown to match those of the real biological phenomenon. A direct comparison with specific cell differentiation processes and the identification of genetic network features that are linked to specific differentiation traits requires the design of a suitable Boolean network such that its dynamics matches a set of target properties. To the best of our knowledge, the only current method for addressing this problem is a random generate and test procedure.

In this work we present an automatic design method for this purpose, based on metaheuristic algorithms. We devised two variants of the method and tested them against random search on typical abstract differentiation trees. Results, although preliminary, show that our technique is far more efficient than both random search and complete enumeration and it is able to find solutions to instances which were not solved by those techniques.

1 Introduction

Cell differentiation is the process whereby a cell undergoes a *cell type* change, from less specialised (e.g. stem cells) to more specialised types (e.g. neurons). This process is characterised by highly complex dynamics, being the result of the interactions among genes and possibly other molecular agents. Cell differentiation seems to be involved also in the development of complex diseases, such as cancer, and computational models of this process may help understand the dynamics of such diseases.

Recently, a cell differentiation model based on Boolean networks [1,2] subject to noise has been proposed [3,4]. In this work, we address the problem of finding

© Springer International Publishing AG 2017
F. Rossi et al. (Eds.): WIVACE 2016, CCIS 708, pp. 91–102, 2017.
DOI: 10.1007/978-3-319-57711-1_8

a Boolean network such that its dynamics generates a given differentiation tree.[1] This problem is faced by casting the search of Boolean network functions into an optimisation problem, in which the objective function quantifies the match between the differentiation tree originated by the Boolean network dynamics and the target one. The optimisation problem is tackled by metaheuristic algorithms, which seem particularly suitable for this kind of problems, as they trade the proof of optimality for efficiency and make it possible to solve problems characterised by a wide search space [5]. Indeed, the model is intrinsically stochastic, so the evaluation of a candidate solution is affected by error and therefore a proof of optimality is useless. This approach will enable us to study the common properties of Boolean networks able to express specific differentiation dynamics features with the aim of identifying generic properties linking genetic network with differentiation tree features.

The paper is organised as follows. In Sect. 2 the differentiation model is introduced. In Sect. 3 the automatic design process is presented and along with the algorithms developed. Results are shown and discussed in Sect. 4 and future works are outlined in Sect. 5.

2   TES Differentiation Model

The cell differentiation model we consider in this work is based on Boolean networks subject to noise and it has been presented in [3, 4].

This model reproduces the main abstract properties of cell differentiation, which are:

1. *Different degrees of differentiation*: totipotent stem cells can give rise to any cell type; pluripotent and multipotent cells can give rise to several, but not all, cell types.
2. *Stochastic differentiation*: a population of identical multipotent cells can generate different cell types, in a stochastic way.
3. *Deterministic differentiation*: specific signals can trigger the development of a multipotent cell into a well-defined type; signals correspond to the activation or deactivation of selected genes or groups of genes.
4. *Limited reversibility*: the differentiation process is almost always irreversible, but there are limited exceptions in which a cell can come back to a previous stage under the action of appropriate signals.
5. *Induced pluripotency*: fully differentiated cells can come back to a pluripotent state by modifying the expression level of some genes.
6. *Induced change of cell type*: the expression of few transcription factors can convert one cell type into another.

While we refer to the original works [3, 4] for more details, let us recall that the model is based on Boolean networks (BNs), which are a prominent example of

---

[1] In general, the so-called lineage tree may not be a proper tree structure, but rather a graph. However, without loss of generality, in this work we will focus on tree structures.

complex dynamical systems and they have been introduced by Kauffman [1,2] as a genetic regulatory network (GRN) model. A BN is a discrete-state and discrete-time dynamical system whose structure is defined by a directed graph of N nodes, each associated to a Boolean variable x_i, $i = 1, \ldots, N$, and a Boolean function $f_i(x_{i_1}, \ldots, x_{i_{K_i}})$, where K_i is the number of inputs of node i. The arguments of the Boolean function f_i are the values of the nodes whose outgoing arcs are connected to node i. The state of the system at time $t, t \in \mathbb{N}$, is defined by the array of the N Boolean variable values at time t: $s(t) \equiv (x_1(t), \ldots, x_N(t))$. The most studied BN models are characterised by *synchronous* dynamics and *deterministic* functions. In this setting, given an initial condition, the dynamics of the networks can be described by means of a unique *trajectory*, which is a sequence of states at consecutive time instants. The asymptotic states of (finite) systems with such a dynamics are called *state attractors* (or simply *attractors*), which are cyclic repetitions of a sequence of network states.

The differentiation model under study considers BNs subject to noise, such as the transient flip of a node value. Attractors of BNs are unstable with respect to noise even at low levels. In fact, even if the flips last for a single time step one sometimes observes transitions from the original attractor to another one. Here we are interested in the asymptotic behaviours of a noisy BN. In such networks, in general the higher this noise, the higher the probability to move across attractors. High levels of noise correspond to pluripotent cell states, where the BN trajectory can wander freely among the attractors; conversely, low levels of noise induce low probabilities to jump between attractors, thus representing the case of specialised cells. A fundamental role in the model is played by *threshold ergodic sets* (TES$_\theta$) which are sets of attractors in which the dynamics of the network remains trapped, under the hypothesis that attractor transitions with probability less than threshold θ are not feasible.[2] The transitions between attractors and their probabilities are summarised in the *attractor transition matrix* (ATM). Thresholds are a dual concept w.r.t. noise: in general, high levels of noise allow most transitions, while low levels enable only a few. Conversely, low threshold values allow most transitions, while high values cut a large number of edges in the ATM.

3   Automatic Design of BN with Predefined TES Tree Structures

The generic abstract properties of the model have been already shown to match those of the real biological phenomenon. A direct comparison with specific cell differentiation processes would require to design a BN (i.e. topology and node transition functions) such that its dynamics gives origin to a differentiation tree matching the properties of the real case at hand. The BN differentiation tree is characterised by the attractor set of the BN and the transitions between them, as

---

[2] This hypothesis is supported by the observation that cells has a finite lifetime, which enables their dynamics to explore only a portion of the possible attractor transitions.

well as their probabilities. Not surprisingly, attaining such a complex dynamics by designing a BN by hand is not possible and an approach based on brute force is definitely impractical; indeed, the number of N nodes networks with exactly k inputs per node is $(2^{2^k})^N$. Notably, each candidate solution, i.e. a BN, is evaluated by computing its ATM, which is a highly demanding computational operation. Therefore, an automatic design method able to efficiently explore the search space is required. To the best of our knowledge, the only current method for attempting to attack this problem is a random generate and test procedure [6], which draws BNs at random until either an acceptable solution is found or the time limit is reached.

In this work we present an automatic design method for this purpose, based on metaheuristic algorithms [5]. This approach maps the BN design into an optimisation problem, where functions and topology of the BN are considered as decision variables and a measure of the matching between the BN differentiation tree generated by its ATM and a target differentiation tree is used as objective function. The objective function we defined for our algorithms is a combination of two tree distance measures: the *edit distance*, E, and the *histogram distance*, H (both distance measures have been mentioned in [6]). The tree edit distance between two trees is the minimum cost sequence of node edit operations (node deletion, node insertion, node rename) that transforms one tree into the other[3]. The histogram distance is a similarity measure between the current tree (C) and the desired tree (D), and is defined as:

$$d = \sum_{l=0}^{l^*} \sum_{k=0}^{k^*} \mid n_C(k,l) - n_D(k,l) \mid \tag{1}$$

where l^* denotes the maximum depth and k^* the maximum number of children nodes in both trees. The function $n_C(k,l)$ computes the number of nodes at the level l with k children in the current tree, and $n_D(k,l)$ respectively for the desired tree [6]. In this way the histogram distance gives us a measure of the structural similarity, level by level, between the two trees; obviously, the lower the histogram distance is, the more similar two trees are. However the histogram distance might result null even if the two trees in exam are different: this may occur because this measure takes into consideration one level at a time. Several combinations of the two distances have been tested; the one leading to the best results is $F = E + (E \times H)$, which was used for the final experiments. The intuition supporting the success of this combined function is that the product between E and H initially prevails and guides the search towards regions of the landscape characterised by differentiation trees close to the target one; once this product becomes negligible, the search is then guided by E and refines the solution. A thorough landscape analysis, which would provide insights on the effectiveness of this specific combination, is subject of future work.

We devised two variants of the method, each based on a different metaheuristic algorithm; a simpler one is based on *adaptive walk*, designed mainly for test

---

[3] http://tree-edit-distance.dbresearch.uni-salzburg.at/.

Algorithm 1. Adaptive Walk

Input: N number of nodes, K incoming degree for each node, p bias, *thresholds* thresholds list, *searchTree* desired tree, *maxIterations* number of the maximum iterations.

```
 1: bn ← GENERATERANDOMNETWORK(N, K, p)
 2: bestNetwork ← bn
 3: tesTree ← CREATETESTREE(bn, thresholds)
 4: distance ← COMPUTEDISTANCE(tesTree, searchTree)
 5: i ← 0
 6: while i < maxIterations & distance > 0 do
 7:     randomFlip ← GENERATEFLIP()
 8:     bn ← MODIFYNETWORK(bn, randomFlip)
 9:     tesTree ← CREATETESTREE(bn, thresholds)
10:     newDistance ← COMPUTEDISTANCE(tesTree, searchTree)
11:     if newDistance > distance then
12:         bn ← MODIFYNETWORK(bn, randomFlip)
13:     else
14:         distance ← newDistance
15:         bestNetwork ← bn
16:     end if
17:     i ← i + 1
18: end while
19: return bestNetwork
```

purposes and a more advanced one is implemented according to a strategy called *variable neighbourhood search*, which is capable of efficiently exploring the search space and escaping from local minima. It is important to stress that a BN whose ATM can be used to obtain a given target differentiation tree just represents one possible model for the real system to be matched. For this reason, randomised techniques are of great help as they make it possible to explore different solutions and provide an ensemble of hypotheses. To this aim, (stochastic) metaheuristic methods are indeed particularly effective as they can be easily adapted so as to provide a wide coverage of the solutions space, e.g. by penalising already visited search space areas or by defining proper re-initialisation mechanisms that make use of some sort of memory so as to start the new search from search space areas not yet explored.

As a first step, we devise algorithms to search in the space of Boolean functions, keeping the topology of BNs constant. We consider BNs with exactly k inputs per node with random topology (without self-arcs). As experimentally shown in [7], this choice is not restrictive.

In the following, we illustrate the search algorithms. For more detail see [8].

3.1   Adaptive Walk Algorithm

The AW algorithm (see Algorithm 1) performs a stochastic descent: it starts from a randomly generated BN and after the execution of each move the resulting solution is accepted if it is not worse—w.r.t. the objective function—than the current solution. A move consists in a flip, from 0 to 1 or vice versa, of a random entry in the truth table of a randomly chosen node. So, a flip changes the genome of the gene regulatory network since it modifies the Boolean function of a node and therefore the response of a gene to certain stimuli.

Observe that this algorithm allows moves that produce solutions with values of the objective function equal to the current one, called *sideways moves*. In this way, the search is able to explore search landscape plateaus. From a biological modelling point of view, sideways moves accomplish the possibility of exploring path in the search space composed of neutral networks, i.e. different networks with the same objective function evaluation, a concept also related to genetic robustness.

The algorithm has been also optimised with a bit of memory: in order to avoid the repeated evaluation of the same network, we forbid to repeat the flip of the previous step.

The search process terminates when the objective function reaches zero (the differentiation tree found corresponds to the target one) or when the number of maximum iterations is reached.

3.2   VNS-Like Algorithm

The second algorithm we present is a metaheuristic technique inspired by *Variable Neighbourhood Search* (VNS).[4] This algorithm is a variant of the previously presented algorithm. AW starts with a randomly chosen network and applies an intensification strategy by making a flip to one output entry at a time. However in this way the search process, depending on the starting solution, might get trapped into local minima with no possibilities to escape or into areas of the search landscape that does not contain "good" quality solutions. For this reason we have added a diversification strategy to our algorithm. The process of diversification is implemented by increasing the number of flips if the search process does not find a solution better than the current one for a given number of steps. A better solution corresponds to a BN able to express a differentiation dynamics more similar to the desired one, i.e. with a lower value of objective function than to the one obtained by the current network. Increasing the number of random flips helps the search process to escape from local minima and it is similar to the change of neighbourhood in case of no improvements that is present in the classical VNS. As soon as a better solution is found, the number of flips is brought back to 1 and so the intensification process restarts until the objective function reaches zero or the number of maximum iterations is reached. When the number of flips is equal to 1, this algorithm behaves exactly like AW.

See Algorithm 2 for a pseudocode description of the VNS algorithm.

---

[4] See [5,9] for details on this technique.

Algorithm 2. Variable Neighbourhood Search

Input: N number of nodes, K incoming degree for each node, p bias, *thresholds* thresholds list, *searchTree* desired tree, *maxIterations* number of the maximum iterations, *maxNoImpovement* number of iterations max without improvements.

```
1:  bn ← GENERATERANDOMNETWORK(N, K, p)
2:  bestNetwork ← bn
3:  tesTree ← CREATETESTREE(bn, thresholds)
4:  distance ← COMPUTEDISTANCE(tesTree, searchTree)
5:  noImprovement ← 0
6:  numFlip ← 1
7:  i ← 0
8:  while i < maxIterations & distance > 0 do
9:      if noImprovement = maxNoImpovement then
10:         noImprovement ← 0
11:         numFlip ← numFlip + 1
12:         if numFlip > N then
13:             return bestNetwork
14:         end if
15:     end if
16:     randomFlips ← GENERATEFLIPS(numFlip)
17:     bn ← MODIFYNETWORK(bn, randomFlips)
18:     tesTree ← CREATETESTREE(bn, thresholds)
19:     newDistance ← COMPUTEDISTANCE(tesTree, searchTree)
20:     if newDistance > distance then
21:         bn ← MODIFYNETWORK(bn, randomFlips)
22:     else
23:         distance ← newDistance
24:         bestNetwork ← bn
25:         if newDistance = distance then
26:             noImprovement ← noImprovement + 1
27:         else
28:             noImprovement ← 0
29:             numFlip ← 1
30:         end if
31:     end if
32:     i ← i + 1
33: end while
34: return bestNetwork
```

4 Results

We evaluated the performance of AW and VNS on target differentiation trees that were defined on the basis of common differentiation tree features, such as the hematopoietic lineage [10]. We defined nine different tree structures, trying to capture the main features. Target trees are depicted in Fig. 1. The threshold values at which the TESs were split have been chosen so as to be distributed in the interval $[0, 1]$ so as to capture changes in the differentiation tree

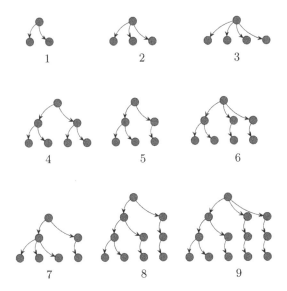

Fig. 1. Differentiation tree structures used as target for the search process.

corresponding to significantly different noise levels, and also to require a high probability to return to the same TESs when at a leaf of the tree. For comparison purposes, thresholds have been set to 0.2, 0.4, 0.6. The evaluation of cases with different choices for the threshold values is subject of ongoing work.

As a baseline comparison, we also run a random search that simply generates random BNs. This algorithm was allowed to generated as many networks as the maximum number of evaluations allowed to AW and VNS. For trees 1 to 7, a maximum of 10^4 evaluations has been allowed, while for trees 8 and 9, which are deeper than the previous ones, we set the maximum number of evaluations to 5×10^4.

Experiments were run on 10 nodes BNs with $k = 2$ and with $k = 3$. Initial RBNs with $k = 2$ were generated with Boolean function bias equal to 0.5, so as to start with BNs in the so-called critical regime [11]. Conversely, initial RBNs with $k = 3$ were generated so that ordered, chaotic and critical regions were sampled. More precisely, we generated networks with Boolean function bias equal to 0.1, 0.5 and 0.79, respectively.

As algorithms are stochastic, 30 independent runs for each case were run and statistics were collected in case the algorithm attained at least 20 successes out of 30. Results are shown in Figs. 2 and 3 by means of boxplots, which provide a visual representation of the distributions.

Trees 1, 2 and 3 are quite trivial, as they are composed of a root and some children nodes. Results on these trees are qualitatively similar: AW and VNS perform better than Random and they are equivalent in terms of number of iterations. It is interesting to observe that if initial solutions are sampled in the ordered regime, more iterations are required to design the target BNs; this

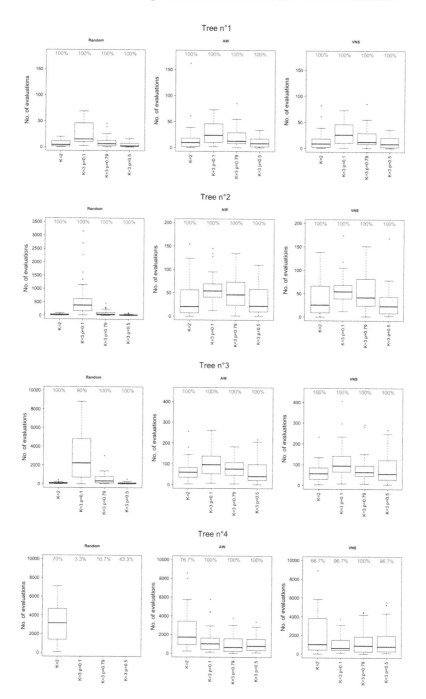

Fig. 2. Boxplot summarising the results for differentiation trees 1 to 4. Boxplots are drawn only in those cases in which the algorithm attained at least a success ratio of 20/30. The success ratio is reported at the top of the plots.

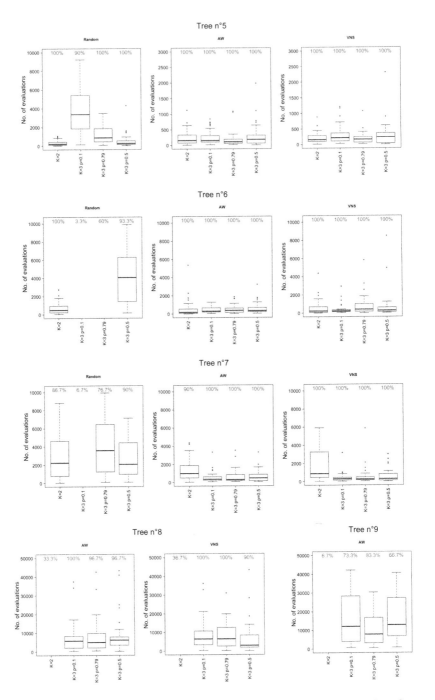

Fig. 3. Boxplot summarising the results for differentiation trees 5 to 9. Boxplots are drawn only in those cases in which the algorithm attained at least a success ratio of 20/30. The success ratio is reported at the top of the plots.

fact can be explained by considering that ordered BNs have typically very few attractors, so the search has first to find networks with a sufficient number of attractors and then to modify the attractors landscape.

Trees 4 to 7 have depth 2 and are generally more difficult to be obtained, and the depth 3 of trees 8 and 9 are even more demanding. We first observe that random search is much less effective than AW and VNS, if not completely unable to find the target tree. Moreover, we note that there is no clear winner between AW and VNS. This result is a bit surprising, as we would expect VNS to be superior to AW; we conjecture that the cause of this behaviour has to be found in the structure of the search landscape, which is likely to be quite uncorrelated, which makes gradual exploration search strategies quite ineffective. An investigation of the search landscape is subject of future work.

5   Conclusion and Future Work

The techniques presented have shown to be superior to random search and able to robustly find BNs matching target differentiation trees with different characteristics. However, this work is just the first step towards the development of efficient techniques for the automatic design of BNs for cell differentiation. Since the computation of the ATM is the most costly computational operation in this process, we are trying to improve the methods so as to reduce the number of evaluations as much as possible. This may be achieved, for example, by introducing heuristics in the choice of the local moves, with the aim of performing an evaluation only for relevant moves. To tackle this problem, further metaheuristic algorithms can also be used besides AW and VNS. For example, one may want to use evolutionary computation techniques or also model-based search methods [12], which may be adapted so as to provide an estimation of the probability of finding a solution. Moreover, the link between differentiation tree structure and search landscape (of course depending on the objective function) has to be investigated. In fact, information on the properties of the landscape may be used to choose the solver most suited for a given tree. The approach presented in this work will also enable us to identify common features among the BNs able to produce some biological plausible differentiation trees, with the aim to find generic properties in gene regulatory networks of real organisms. In addition, following the *ensemble approach* proposed by Kauffman [13] we can generate and study different network instances and detect the properties of the ensemble that shows statistical features that match those of real cells. The techniques we propose are particularly suitable for this task as they perform a guided sampling in the BN search space and they are more efficient than random search. Finally, this approach may be extended by adding specific constraints motivated by biological plausibility, such as forcing specific activation patterns among genes.

Acknowledgements. The authors thank Alex Graudenzi and Chiara Damiani for helpful discussions and suggestions.

References

1. Kauffman, S.A.: Metabolic stability and epigenesis in randomly constructed genetic nets. J. Theor. Biol. **22**(3), 437–467 (1969)
2. Kauffman, S.A.: The origins of order. Oxford University Press, New York (1993)
3. Serra, R., Villani, M., Barbieri, A., Kauffman, S.A., Colacci, A.: On the dynamics of random Boolean networks subject to noise: attractors, ergodic sets and cell types. J. Theor. Biol. **265**(2), 185–193 (2010)
4. Villani, M., Barbieri, A., Serra, R.: A dynamical model of genetic networks for cell differentiation. PLoS ONE **6**(3), e17703 (2011)
5. Blum, C., Roli, A.: Metaheuristics in combinatorial optimization: overview and conceptual comparison. ACM Comput. Surv. **35**(3), 268–308 (2003)
6. Paroni, A., Graudenzi, A., Caravagna, G., Damiani, C., Mauri, G., Antoniotti, M.: CABeRNET: a cytoscape app for augmented boolean models of gene regulatory NETworks. BMC Bioinformatics **17**, 64–75 (2016)
7. Benedettini, S., Villani, M., Roli, A., Serra, R., Manfroni, M., Gagliardi, A., Pinciroli, C., Birattari, M.: Dynamical regimes and learning properties of evolved boolean networks. Neurocomputing **99**, 111–123 (2013)
8. Braccini, M.: Automatic design of boolean networks for modelling differentiation trees. Master's thesis, Corso di Studio in Ingegneria e scienze informatiche - Cesena, Università di Bologna (2016)
9. Hansen, P., Mladenović, N.: Variable neighborhood search: principles and applications. Eur. J. Oper. Res. **130**, 449–467 (2001)
10. Alberts, B., Johnson, A., Lewis, J., Raff, M., Roberts, K., Walter, P.: Molecular Biology of the Cell, Chapt. 23. Garland Science, 5th edn. (2007)
11. Bastolla, U., Parisi, G.: A numerical study of the critical line of kauffman networks. J. Theor. Biol. **187**(1), 117–133 (1997)
12. Zlochin, M., Birattari, M., Meuleau, N., Dorigo, M.: Model-based search for combinatorial optimization: a critical survey. Ann. Oper. Res. **131**(1), 373–395 (2004)
13. Kauffman, S.A.: A proposal for using the ensemble approach to understand genetic regulatory networks. J. Theor. Biol. **230**(4), 581–590 (2004)

Model-Based Lead Molecule Design

Alessandro Giovannelli[1], Debora Slanzi[1,2(✉)], Marina Khoroshiltseva[1],
and Irene Poli[1,2]

[1] European Centre for Living Technology, S. Marco 2940, 30124 Venice, Italy
{alessandro.giovannelli,debora.slanzi,
marina.khoroshiltseva,irenpoli}@unive.it
[2] Department of Environmental Sciences, Informatics and Statistics,
Ca' Foscari University of Venice, via Torino 155, Mestre, Italy

Abstract. "Lead molecule" is a chemical compound deemed as a good
candidate for drug discovery. Designing a lead molecule for optimization involves a complex phase in which researchers look for compounds
that satisfy pharmaceutical properties and can then be investigated for
drug development and clinical trials. Finding the optimal lead molecule
is a hard problem that commonly requires searching in high dimensional
and large experimental spaces. In this paper we propose to discover the
optimal lead molecule by developing an evolutionary model-based approach where different classes of statistical models can achieve relevant
information. The analysis is conducted comparing two different chemical representations of molecules: the amino-boronic acid representation
and the chemical fragment representation. To deal with the high dimensionality of the fragment representation we adopt the Formal Concept
Analysis and we then derive the evolutionary path on a reduced number
of fragments. This approach has been tested on a particular data set of
2500 molecules and the achieved results show the very good performance
of this strategy.

Keywords: Fragment-based lead discovery · Formal Concept Analysis ·
Evolutionary experimental design

1 Introduction

In the research field of drug discovery a key problem is the design of functional molecules that affect proteins associated with diseases. This issue is usually
referred to as drug design, and the use of computer-aided methods can help in
the difficult process of the discovery and the optimization of the lead molecules,
by eliminating compounds with poor capacity to satisfy the essential properties
of a drug, such as Absorption, Distribution, Metabolism, Excretion and Toxi-
city (ADMET) and by selecting the most promising candidate solutions. This
optimization is generally an hard problem since extremely large experimental
spaces have to be analyzed. With the aim of developing a procedure able to
discover the optimal molecule, several approaches have been proposed over the

© Springer International Publishing AG 2017
F. Rossi et al. (Eds.): WIVACE 2016, CCIS 708, pp. 103–113, 2017.
DOI: 10.1007/978-3-319-57711-1_9

last few years mostly based on computational search and stochastic optimization methods; in particular, nature-inspired algorithms such as genetic algorithms or artificial neural networks have been quite successful in this task [1,2]. Addressing the optimization with an evolutionary approach, we recently developed an experimental design where the evolution is driven by predictive statistical models [3–5]. This model-based Evolutionary Design for Optimization (EDO) approach leads to a drug design that is sequential, adaptive and self-organizing.

In this paper we address the problem of designing and modeling experimental data with the aim of discovering optimal molecules. In particular we study an experimental space where molecules have a chemical fragment representation. Usually the number of fragments in molecules is extremely large (several thousands) and this generates the problem of high dimensionality in modeling the experimental data. To reduce the dimensionality of data, we adopt the approach of Formal Concept Analysis, FCA hereafter [6–9]. FCA is a mathematical procedure to data analysis which derives concepts of hierarchies from a collection of objects and attributes. Specifically, formal concepts are sets of objects that share a defined subset of attributes. FCA derives these sets of objects by building *lattice nodes* where each node specifies relationships among objects and attributes.

In the drug discovery context, molecules can be considered as objects and the fragments composing the molecules as attributes. Therefore FCA produces lattice nodes which correspond to clusters of molecules sharing common fragments. Selecting only the common fragments in the lattice nodes which represent specific molecules for the problem under study will then reduce the high dimensionality of the experimental space.

To evaluate the optimization procedure based on EDO design and FCA we consider the dataset consisting of 2500 molecules reported in [1]. This dataset represents the test set on which we can evaluate the performance of the optimization procedure. The goal of the research is in fact to discover the optimal lead molecule using a very small set of experimental points (less than 5% of the whole experimental space).

In this paper we develop the procedure of model-based EDO approach by estimating the following models: Lasso approach [10], Stepwise regression [11] and Boosting regression [12].

The performance of the approach is evaluated by comparing the different formulations of model-based EDO with the basic evolutionary procedure in the form of genetic algorithm for optimization (GAO). The results are then compared in terms of two different chemical structure representations of the molecules: the amino-boronic acid composition $(A_i B_j)$ and the fragment composition reduced by FCA. The analysis shows a better performance of model-based EDO approach with respect to GAO approach and also a better performance of fragment FCA representation with respect to amino-boronic acid representation.

The paper is organized as follows. In Sect. 2 we provide a description of the data analyzed in the study. Section 3 we introduce the EDO design suited for lead molecule optimization with fragment representation and in Sect. 4 we provide the main results achieved in the optimization of the target of the experimentation. In Sect. 5 we provide some conclusions.

2   Data Description

Data are concerned with a set of molecules described by a response variable, namely pIC50 MMP12 - Activity, and a set of explanatory variables initially in the form of amino-boronic acid compounds (A_iB_j) and later in the form of fragments (with different sizes). More specifically in the first representation, we consider a dataset of 2500 molecules described by the amino acid compound A_i and the boronic acid compound B_j, with $i, j = 1, ..., 50$, as presented in [1]. This experimental space, consisting of 2500 points with A_iB_j molecule representation, is defined as $\boldsymbol{\Omega_1}$.

From this dataset we derive a successive different representation of the molecules by considering the fragment representation. The initial full set of fragments describing the 2500 molecules, consists of 22272 different fragments. Starting from this dataset we conduct FCA analysis and we achieve 4059 lattice nodes.

Each lattice node, say C_k with $k = 1, ..., 4059$, is composed by the subset of molecules A_iB_j that share a subset of fragments $F_m \subset F$, where F is the full set of 22272 fragments. From FCA analysis we can identify the fragment composition for each separate compound A_i and B_j. To achieve this result, we identify 100 nodes of 4059, one for each possible A_i and B_j, with the following representation

$$C_{Ai} = \langle \{A_iB_1, \ldots, A_iB_{50}\}, \{F_{Ai}\}\rangle, \text{ with } i = 1, \ldots, 50$$
$$C_{Bj} = \langle \{A_1B_j, \ldots, A_{50}B_j\}, \{F_{Bj}\}\rangle, \text{ with } j = 1, \ldots, 50$$

where $F_{Ai}, F_{Bj} \subset F$. Then taking the union of the generic F_{Ai} with F_{Bj}, we get a new experimental space consisting of 2500 points described by 957 fragments, where these selected fragments are those relative to each compound A_i and B_j. With this fragment representation, each molecule of the experimental space can be defined as

$$S_{ABm} = \langle \{A_iB_j\}, \{F_{Ai} \bigcup F_{Bj}\}\rangle, \text{ with } i, j = 1, \ldots, 50 \text{ and } m = 1, \ldots, 2500.$$

Proceeding in the analysis, we then discover the presence of linear dependences (multicollinearity) among some fragments in the dataset: there are groups of specific fragments that appear (or not appear) always together in the composition of the compounds. This allows us a further reduction of the dataset by excluding the linearly dependent fragments and considering just one representative fragment. Applying this procedure we reduce the dataset from 957 to 175 fragments. This dataset $\boldsymbol{\Omega_2}$ will be then used in the analysis.

Therefore with the goal of evaluating the role of different chemical representations in the optimization process, we develop the analysis on the two following datasets:

1. Dataset Ω_1, consisting of 2500 points with A_iB_j molecular representation where each point is described by 2 categorical variables with 50 levels each.
2. Dataset Ω_2, consisting of 2500 points where each point has the fragment representation of the compounds A_i and B_j achieved by FCA analysis and multicollinearity reduction. Each point of Ω_2 is then described by dummy variables $(X_{dm}, d = 1, \ldots, 175$ and $m = 1, \ldots, 2500)$, indicating the presence/absence of fragments in the particular molecule

$$X_{dm} = \begin{cases} 1 & \text{if } S_{ABm} \text{ is in the molecule } A_iB_j \\ 0 & \text{otherwise} \end{cases}$$

In Fig. 1 we report the fragment representation of the whole experimental space. In particular, in Fig. 1(a) we represent the data molecular structure when all the 22272 original fragments are considered; whereas in Fig. 1(b) we represent the data molecular structure when only the 175 fragments selected from FCA are reported.

3 The Model-Based Evolutionary Design for Optimization

Addressing high dimensional optimization problems, the Evolutionary Design for Optimization (EDO) approach was recently proposed and successfully developed in several biochemical applications [3–5]. EDO approach selects a very small initial set of candidate solutions, tests them and achieves a first set of experimental responses. These data are then used to estimate predictive statistical models that yield information on the most promising candidates or hypotheses suggesting new experimental tests. The process is iteratively repeated, generation after generation, maintaining the same size in each population of experimental points and ends when the optimum value is achieved, or the maximum total number of experimental points is reached. The model-based design is evolutionary and adaptive, since it can be constructed using different classes of models in each generation depending on the data resulting from experimentation. In this paper we focus on some classes of statistical models suited for high dimensional data and we derive the following approaches:

- EDO-Lasso, where the statistical model is based on Least Absolute Shrinkage and Selection Operator [10];
- EDO-Stepwise, where the statistical model is derived by estimating a forward Stepwise regression [11];
- EDO-Boosting, where the statistical model is based on L_2-Boosting regression model [12].

Following [1], the structure of EDO design to discover the lead molecule consists of randomly selecting an initial small population, in this study 20 molecules, and then evaluating the response variable value of each molecule. We select as response variable of the experimentation the molecular Activity. In this study the values of

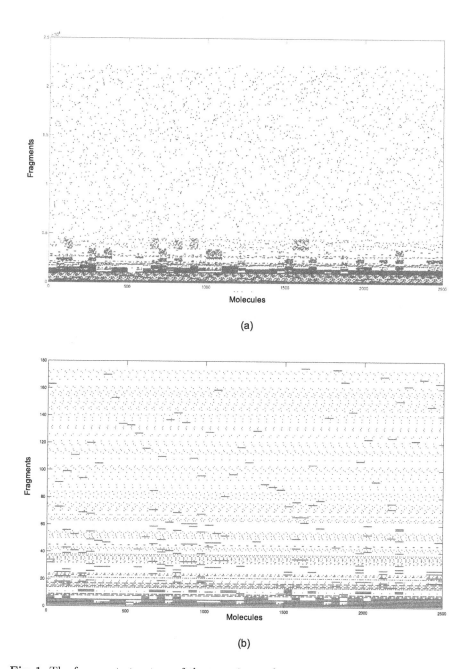

Fig. 1. The fragment structure of the experimental space consisting of 2500 molecules is described in (a) with the whole set of 22272 fragments and in (b) with the selected FCA set of 175 fragments.

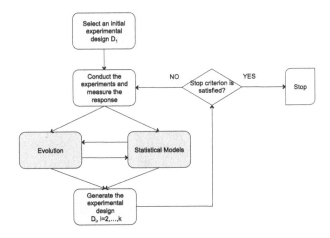

Fig. 2. The EDO approach

this response variable are achieved directly from the dataset provided by [1]. With the goal of maximizing the Activity level, we then estimate and evaluate statistical models and achieve the Activity predicted values on the experimental space. In the evolutionary algorithm, the next population of experiments is then built by selecting the 20 molecules with the best predicted response values. This new set of candidate points is then evaluated and the process is repeated for 7 generations to reach a total of 140 compositions. A graphical representation of EDO approach is presented in Fig. 2.

4   The Optimization of the Activity Response Variable

We present the performance of the optimization process by developing a comparison of the results achieved by using the original A_iB_j molecular representation, i.e. Ω_1, and the results achieved by FCA fragment representation Ω_2. We run 1000 Monte Carlo simulations to study the robustness of the procedures with respect to changes in the initial population. The results are analyzed and compared both with the genetic algorithm optimization (GAO) and EDO approach constructed with the estimation of three different statistical models previously introduced. In Table 1 we report the proportion the best Activity value, i.e. $Activity = 8$, and the proportion of values in the region of optimality for the system, i.e. $Activity \geq 7.5$, achieved by the procedures in 1000 runs. With dataset Ω_1, namely with the amino-boronic representation, EDO approach shows much better performances achieving the maximum Activity value in the 83.5% of the runs with respect to the 21.5% of the GAO. Considering the region of optimality, for Activity values grater than 7.5, both the EDO and GAO can reach optimal performances (100% of EDO-Boosting and 99% of GAO). With dataset Ω_2, namely with the fragment representation, we notice again a better

Table 1. Percentage of best Activity values in 1000 runs.

	A_iB_j dataset			FCA fragment dataset	
	Activity = 8	Activity ≥ 7.5		Activity = 8	Activity ≥ 7.5
GAO	21.5%	99.0%	GAO	15.0%	86.7%
EDO-Lasso	76.4%	99.3%	EDO-Lasso	84.4%	100.0%
EDO-Stepwise	63.9%	99.4%	EDO-Stepwise	78.2%	99.5%
EDO-Boosting	83.5%	100.0%	EDO-Boosting	82.7%	99.3%

performance of EDO approach with respect to GAO where EDO-Lasso reaches the maximum Activity value in 84.4% of 1000 runs while the GAO reaches the maximum Activity value just in the 15% of the runs. Also for the optimality region, we notice the very good performance of EDO-Lasso reaching values in the best region in the 100% of the times with respect to GAO than reaches values in this region in 86.7% of the runs. We also notice that EDO-Lasso and EDO-Stepwise approaches give much better performances with the fragment representation with respect to the amino-boronic acid representation showing that a deeper knowledge of the fragment representation can help in optimizing the lead molecule.

In Fig. 3 we present the Monte-Carlo simulation results of the best response values achieved with both evolutionary procedures and for both chemical representations. In particular in this figure we compare the number of runs in which GAO and EDO procedures achieve the best response values when developed for the A_iB_j dataset (Fig. 3(a)) with respect to the FCA fragment dataset (Fig. 3(b)). We highlight the good performances of all EDO procedures in reaching the global optimum in a high number of runs, especially for EDO-Lasso.

For this particular statistical model, we also report the boxplots of the 1000 runs results of the optimization procedure at each generation (Fig. 4). We notice that the EDO-Lasso approach presents very high quartile values starting from the second generation, and also a much smaller size of the box showing a greater precision of the results (smaller variability) when developed for the FCA fragment dataset (Fig. 4(b)) with respect to A_iB_j dataset (Fig. 4(a)). This representation confirms the capacity of EDO procedures to reach the optimum value (8) from the second generation on.

Therefore, in terms of comparison between the two different molecule representations, we can notice that the FCA chemical fragment representation can lead to better results with respect to the A_iB_j amino-boronic acid representation. In this case we also notice that this representation leads to similar results for all the EDO procedures with a much better results with respect to the GAO procedure.

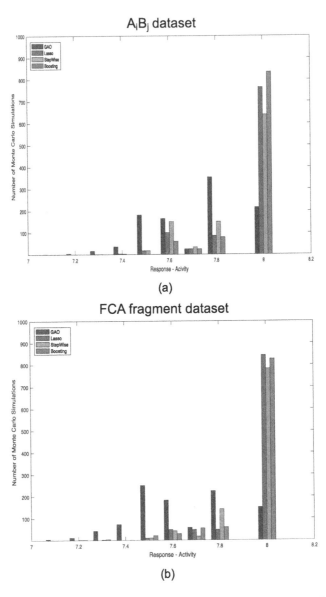

Fig. 3. The frequency distribution of Activity experimental response values achieved in 1000 Monte-Carlo simulations of the optimization procedures with (a) the amino-boronic acids representation and (b) the FCA chemical fragment representation.

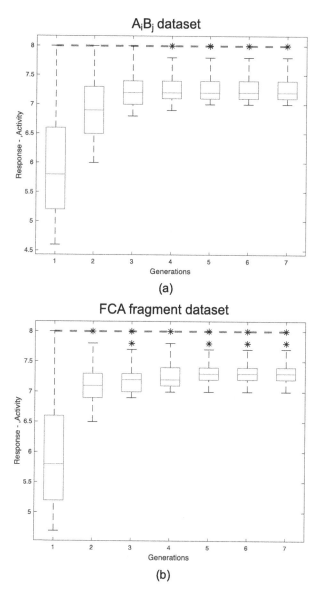

Fig. 4. The boxplots of the Activity experimental values achieved in 1000 runs of EDO-Lasso with (a) the amino-boronic acid representation and (b) the FCA chemical fragment representation.

5   Concluding Remarks

Aim of this research was the development of efficient and effective strategies to construct small molecules under the hypothesis that their high activity levels can affect a particular disease state. In the experimental context, there are several ways to construct these molecules which involve also different chemical representations with different forms of information. Under the assumption that a deep and more informative structure can bring new chemical information, we developed the analysis by building an evolutionary procedure to discover the highest activity molecule and we confront two chemical representations: the combination of selected amino-boronic acids representation and the chemical fragment representation. Given the high dimensional structure of the fragment experimental space, where each molecule is represented by a set of 22272 fragments, a Formal Concept Analysis has been developed and a smaller set of 175 fragments has been identified. The model-based evolutionary strategy has then been conducted with different statistical models and the results show that the structural fragment representation achieved with FCA leads to a more effective optimization with respect both to the amino-boronic acid representation and to GAO procedures.

Acknowledgements. The authors would like to acknowledge Professor Philip J. Brown and the GlaxoSmithKline Medicines Research Centre (UK) for the very fruitful collaboration in developing this research.

References

1. Pickett, S.D., Green, D.V.S., Hunt, D.L., Pardoe, D.A., Hughes, I.: Automated lead optimization of MMP-12 inhibitors using a genetic algorithm. ACS Med. Chem. Lett. **2**(1), 28–33 (2011)
2. Devi, R.V., Siva, S.S., Coumar, M.S.: Evolutionary algorithms for de novo drug design - a survey. Appl. Soft Comput. **27**, 543–552 (2015)
3. Baragona, R., Battaglia, F., Poli, I.: Evolutionary Statistical Procedures: An Evolutionary Computation Approach to Statistical Procedures Designs and Applications. Springer, Heidelberg (2013)
4. Borrotti, M., De March, D., Slanzi, D., Poli, I.: Designing lead optimization of MMP-12 inhibitors. Comput. Math. Methods Med. 1–8 (2014)
5. Slanzi, D., De Lucrezia, D., Poli, I.: Querying Bayesian networks to design experiments with application to 1AGY serine esterase protein engineering. Chemom. Intell. Lab. Syst. **149**, 28–38 (2015)
6. Wille, R.: Restructuring lattice theory: an approach based on hierarchies of concepts. In: Rival, I. (ed.) Ordered Sets. NATO Advanced Study Institutes Series, vol. 83, pp. 445–470. Springer, Dortchet (1982)
7. Kaytoue-Uberall, M., Duplessis, S., Napoli, A.: Using formal concept analysis for the extraction of groups of co-expressed genes. In: Le Thi, H.A., Bouvry, P., Pham Dinh, T. (eds.) MCO 2008. CCIS, vol. 14, pp. 439–449. Springer, Heidelberg (2008). doi:10.1007/978-3-540-87477-5_47
8. Lounkine, E., Auer, J., Bajorath, J.: Formal concept analysis for the identification of molecular fragment combinations specific for active and highly potent compounds. J. Med. Chem. **17**(51), 5342–5348 (2008)
9. Raza, K.: Formal concept analysis for knowledge discovery from biological data. arXiv preprint arXiv:1506.00366 (2015)
10. Tibshirani, R.: Regression shrinkage and selection via the lasso. J. R. Stat. Soc. Ser. B Stat. Methodol. **58**(1), 267–288 (1996)

11. Hastie, T.J., Tibshirani, R.J., Friedman, J.H.: The Elements of Statistical Learning: Data Mining, Inference, and Prediction. Springer Series in Statistics. Springer, New York (2009)
12. Bühlmann, P.: Boosting for high-dimensional linear models. Ann. Stat. **34**(2), 559–583 (2006)

Reducing Dimensionality in Molecular Systems: A Bayesian Non-parametric Approach

Valentina Mameli[1,2]([✉]), Nicola Lunardon[1,2], Marina Khoroshiltseva[1], Debora Slanzi[1,2], and Irene Poli[1,2]

[1] European Centre for Living Technology, S. Marco 2940, 30124 Venice, Italy
{valentina.mameli,nicola.lunardon,marina.khoroshiltseva,
debora.slanzi,irenpoli}@unive.it
[2] Department of Environmental Sciences, Informatics and Statistics,
Ca' Foscari University of Venice, via Torino 155, Mestre, Italy

Abstract. In this paper we present a methodology that can be used to design experiments of complex systems characterized by a huge number of variables. The strategy combines the evolutionary principles with the information provided by statistical models tailored to the problem under consideration. Here, we are concerned with the process of design molecules, which is a quite challenging problem due to the presence of a high number of variables with a binary structure. Recent works on clustering of binary data and variable selection in the high-dimensional setting allow to develop an approach capable of recovering useful information derived from the incorporation of a grouping structure into the model.

Keywords: Bayesian non-parametric clustering · Evolutionary algorithms for optimization · High-dimensional models · Lead molecule optimization · Penalized regression procedures

1 Introduction

In several scientific research fields, we observe an increasing number of really imposing datasets; imposing in size, for the huge number of measurements provided by technological advances; in dimensions, for the very large number of variables that investigators wish to consider in developing their research; and in complexity, for the high level of connectivity among attributes in these large dimensional data spaces. The increase in the size, dimensionality and complexity of datasets poses a challenging problem in discovering patterns and modelling structures. The development of new statistical tools proposed to analyse these data is therefore crucial in contemporary research development. Many proposals are available in literature with the main aim of reducing the dimensionality and selecting the most relevant variables for the problem under study [1,2]. In the research field of drug discovery, we deal with very high-dimensional and complex problems. More specifically, lead molecule optimization concerns the identification of molecules with required properties and the set of variables that affect these

© Springer International Publishing AG 2017
F. Rossi et al. (Eds.): WIVACE 2016, CCIS 708, pp. 114–125, 2017.
DOI: 10.1007/978-3-319-57711-1_10

properties is extremely large. In these systems, the exploration of the experimental space entails experimentation that should to be limited since it requires high investments of resources. The construction of efficient experimental designs can contribute substantially to obtain valid and accurate experimental results at a minimum cost. Several studies have focused on strategies to design experiments inspired by evolution and the information gathered from statistical models in particular when the experimentation is conducted to search for an optimal value [3–6]. One of the great benefits of these model-based evolutionary designs is that they are very flexible as they can be tailored to the problem under study.

In addressing the lead molecule optimization from a statistical perspective, the task is to detect the set of relevant variables able to predict the desired properties of molecules. Aim of this paper is to contribute to the model-based evolutionary procedure by reducing the dimensionality of the experimental space while achieving the essential informative elements affecting the response of the experiments. The proposed strategy hinges on the idea of reducing the dimensionality by firstly grouping fragments into non-overlapping clusters, selecting the most relevant predictors affecting the response variable, and finally integrating the achieved information in the modelling phase. The information collected from the clustering and modelling phases will be then integrated into the evolutionary rules. Clustering of fragments is performed by Bayesian non-parametric strategies which are flexible and computationally efficient tools. For our purposes we focus on the proposal of [7], which is a Bayesian non-parametric clustering approach developed for binary high-dimensional data. The approach assumes that the distribution of clusters arises from a Dirichlet process and variables are generated by a mixture of Bernoulli distributions whose parameters follow a Beta distribution. The methodology allows the Beta distribution associated to each variable to have its own set of parameters that can be updated from data. Furthermore, the approach, in analogy with Bayesian clustering approaches, and in contrast to routinely used clustering methods, dispenses the user from the choice of the number of clusters in advance. The selection of the most relevant explanatory clusters and variables affecting the response of the experimentation is obtained by using a penalized regression procedure. These procedures are suited for handling high-dimensional data and can be tailored to the subject under consideration. In particular, we focus on the proposal of [8], which is capable of identifying relevant variables by exploiting the grouping structure achieved in the clustering phase.

The good performance of the novel procedure will be shown in its capacity to uncover the optimum value using a very limited set of experimental points, avoiding unnecessary experimentation.

The structure of the paper is as follows. In Sect. 2, we describe the drug discovery process and the key ideas of the proposed design. In Sect. 3, we briefly introduce the Bayesian non-parametric approach for clustering binary data and penalized regression procedures. In Sect. 4, we present the results achieved by the proposed procedure for the lead molecule optimization. Some concluding remarks are given in Sect. 5.

2   Designing Molecular Systems

The drug discovery process is concerned with the design of small molecules capable to inhibit or activate a protein function that can influence a certain disease. Traditionally, the most promising candidates, known as the lead molecules, are identified by screening large libraries composed of millions of small molecules [5]. The search of all possible molecules with the desired biological activity entails a challenging problem which depends on the molecular structure complexity. The search concerns in fact the exploration of a very large experimental space where each point is high-dimensional requiring high investment of resources and time to reach the desired target. Specifically, the lead molecule optimization problem motivating this work emerges from the work of [5] subsequently analyzed by [4], where the goal of the experimentation is to explore a complex experimental space by testing only a small set of points. The experimental space consists of 2500 molecules which are characterized by a set of 22272 fragments. Each fragment is coded as a binary variable representing the presence/absence of the fragment in the molecule. The response of the system is a quantitative variable, which measures the biological activity of the molecule. The challenging optimization task concerns the identification of the lead molecules with very high response values by testing only 140 molecules, i.e. 5.6% of the possible experimental points.

2.1   The Model-Based Evolutionary Design for Optimization

There are numerous examples in the drug discovery literature demonstrating that evolutionary algorithms are useful methods to obtain efficient designs; see for instance [9] and references therein. In this work, we consider the design strategy proposed by [3,4,6] which is developed according to the evolutionary paradigm. The procedure incorporates both the ability of the evolutionary approach to explore the whole experimental space with the capacity of statistical models to identify information driving the optimization. In building the evolutionary design for optimisation, namely EDO-design, a first initial population of experimental points is randomly selected from the whole experimental space. This initial set of experimental points is then synthesized in laboratory to provide the biological activity values representing the response of the experimentation. The data are then processed by estimating statistical predictive models and the information gathered from models is then incorporated into the evolutionary rules in order to select the next small set of candidate points. The process is repeated generation after generation until a total amount of 140 experimental points is reached.

In EDO-design a key role is assigned to the statistical model used for predictions: when the number of predictor variables is much larger than the number of observations (the statistical challenging task known as $p \gg n$) the estimated models could be unreliable and not robust in prediction. In this work we present a clustering-based strategy which will be embedded in a penalized regression procedure able to accomplish subset selection in a stable and computationally efficient fashion. Penalized regression procedures are in fact successfully introduced in recent literature especially in genomic and medicine; see for instance [10–12].

3 Statistical Models for Prediction

We consider the linear regression model with p predictors (or covariates) $X_1, \ldots, X_p$

$$y = X\beta + \epsilon$$

where $X = (X_1, \ldots, X_p)$ is an $n \times p$ design matrix, n is the number of observations, $y = (y_1, \ldots, y_n)^T$ is the vector containing the n observations of the response variable, $\epsilon = (\epsilon_1, \ldots, \epsilon_n)^T$ is the error vector and $\beta = (\beta_1, \ldots, \beta_p)$ is the regression vector of p parameters. The regression vector β is unknown and must be estimated from the data. It is known that when the number of covariates (dimension of the system) considerably exceeds the number of observations ($p \gg n$), the estimation of β poses challenging methodological problems. Many methods have been proposed to address this issue and in the two last decades statistical literature has posed particularly attention to penalized regression procedures [1,8,13,14]. Under the sparsity assumption according to which only a limited number of predictors can affect the response variable, these procedures are able to select the most relevant predictors and simultaneously estimate the regression parameters of the model. Recently, there has been some works which hinge on the idea of simultaneously clustering the predictors and selecting the most informative ones within these clusters by using penalized regression procedures; see for example [15] and references therein. Many of these efforts have been motivated by the need to improve on modelling and prediction capabilities of linear models [12].

Motivated by the aforementioned results, we develop a strategy to improve the drug discovery process by reducing the dimensionality and by selecting the most relevant predictors for the problem under study. The proposed strategy consists of three steps. In the first step, we group the predictors into non-overlapping clusters (or groups) using a Bayesian non-parametric approach suitable for binary data [7]; in the second step, we select the most relevant clusters and predictors by using a penalized regression procedure in which the information obtained in the clustering phase is embedded; finally, we integrate the clustering and modelling steps in EDO-design to optimize the response of the drug discovery system. In the following we introduce the most important statistical features which characterize our strategy: the Bayesian non-parametric approach for clustering binary data and the penalized regression procedures. The approach that we proposed will be called Cluster-based Evolutionary Design for Optimization, namely Clu-EDO.

3.1 The Bayesian Non-parametric Approach for Clustering Binary Data

In this work we focus on the Bayesian non-parametric approach for clustering binary data recently proposed by [7]. Bayesian non-parametric methods are proved to be highly flexible and very efficient in dealing with the complexity of the data. Moreover, they are suited for problems where sparse structures are present and when the number of observations is smaller than the number of covariates. The Bayesian non-parametric approach for clustering binary data presents many benefits over classical clustering algorithms, the most notable of which is the ability to infer the number of clusters from data. The algorithm proposed by [7] is build on the idea of grouping together variables which have a similar function or form a similar pattern. Specifically, assume that x_{ij} represents the presence/absence of the j-th fragment (the j-th predictor X_j) in the i-th molecule, i.e. $x_{ij} = 1$ if X_j is present in the i-th molecule and $x_{ij} = 0$ otherwise, for $i = 1, \ldots, n$ and $j = 1, \ldots, p$. Let c_j denote a latent cluster label for X_j, with $c_j = k$ if X_j is allocated to the k-th cluster, $k = 1, \ldots, K$. Here, K denotes the unknown number of non-overlapping clusters with $K \ll p$. The objective of the clustering approach is to associate to each X_j a unique label c_j with $j = 1, \ldots, p$. The approach assumes that the data x_{ij} are independent draws from a mixture of infinite Bernoulli distributions whose parameters p_{jk} are distributed according to a Beta distribution with parameters (a_{jk}, b_{jk}). These parameters represent the prior knowledge about the presence of X_j in the k-th cluster. According to [7], we assume that $a_{jk} = 1$, while b_{jk} is empirically estimated as $b_{jk} = n/\sum_{i=1}^{n} x_{ij}$.

The Bayesian non-parametric specification is completed by assuming that the probability that the fragment j is allocated into cluster k is Π_k, i.e.

$$P(c_j = k) = \Pi_k, \quad \text{with} \quad \sum_{k=1}^{K} \Pi_k = 1.$$

The random variables $\Pi_1, \ldots, \Pi_K$ have a Dirichlet distribution with parameters α and K. The parameter α identifies the concentration parameter of the Dirichlet distribution and controls the number of clusters: larger values of α will tend to lead to many clusters. According to [7], this parameter has been fixed to 1.1 in order to avoid fragmentation of data into many small clusters.

Then a simulated annealing procedure is used to obtain an optimal estimate of cluster labels $\{c_j | j = 1, \ldots, p\}$ and the posterior distribution of the parameters. In the simulated annealing, we use the same initial parameters (temperature and cooling factor) as in [7], while the number of iterations has been set to 50. The Bayesian non-parametric algorithm was performed by setting the maximum number of clusters K to 150.

3.2   Penalized Regression Procedures

Penalized regression procedures, also known as regularized regression methods, are frequently used in high-dimensional problems as they are able to estimate reliable models also when the number of predictors is much larger than the number of observations. According to these procedures, the vector of regression coefficients $\boldsymbol{\beta}$ is estimated by minimizing an objective function $Q(\cdot)$ composed by a loss function $L(\cdot)$ and a penalty function $P(\cdot)$:

$$Q(\boldsymbol{\beta}) = L(\boldsymbol{\beta}|\boldsymbol{y}, \boldsymbol{X}) + P(\boldsymbol{\beta}|\lambda).$$

The loss $L(\cdot)$ measures the discrepancy between the response variable and its prediction. We focus on the least square loss function, namely

$$L(\boldsymbol{\beta}|\boldsymbol{y}, \boldsymbol{X}) = \frac{1}{2n}(y - X\boldsymbol{\beta})^T(y - X\boldsymbol{\beta}).$$

The penalty function $P(\cdot)$ encourages sparsity and avoids over-fitting in high-dimensional dataset. The parameter λ is a trade-off between the loss and the penalty. The choice of the parameter λ could be addressed by considering the literature about cross-validation or information criteria such as the Akaike information criterion and the Bayesian information criterion. Various penalties have been proposed in the literature, for a recent review see [8]. Among them, we mention the least absolute shrinkage selection operator (lasso) proposed by [13] which is based on the penalty $P(\boldsymbol{\beta}|\lambda) = \lambda \sum_{j=1}^{p} |\beta_j|$. The lasso is widely used in the context of high-dimensional models for variable selection (or dimension reduction). The choice of the penalty function should be tailored to the subject under consideration. For instance in high-dimensional regression settings, choices relating to cluster structures of predictors can be useful for reducing the dimensionality of the system [8]. The information contained in the cluster structure can in fact be exploited in the regression analysis in order to enhance the prediction capacities of models. In these cases, the penalty should provide insight into which are the most informative clusters as well as which are the most relevant variables affecting the response variable within these clusters [10, 12]. Specifically, let suppose that the p predictors can be naturally divided into K non-overlapping clusters. The linear regression model can be written as

$$y = \sum_{k=1}^{K} \tilde{X}_k \tilde{\beta}_k + \epsilon,$$

where $\tilde{X}_k$ is the $n \times d_k$ design matrix formed by the d_k predictors belonging to the k-th cluster, $\tilde{\beta}_k = (\beta_{k1}, \ldots, \beta_{kd_k}) \in \mathbb{R}^{d_k}$ is the vector of regression coefficients of the k-th cluster and ϵ is the error vector.

In this paper we adopt the composite minimax concave penalty whose expression is [8]

$$P(\boldsymbol{\beta}|\lambda) = \sum_{k=1}^{K} \rho_{\lambda, \gamma_O} \left(\sum_{j=1}^{d_k} \rho_{\lambda, \gamma_I} \left(|\beta_{kj}| \right) \right)$$

where ρ_{λ, γ_O} and ρ_{λ, γ_I} are the outer and inner penalties, respectively, able to identify relevant variables by exploiting the information contained in the cluster structure. The parameters γ_O and γ_I are tuning parameters with $\gamma_O = d_k \gamma_I \lambda/2$. For the choice of these parameters we refer to [8] and references therein. The penalty ρ is defined as in [14] and assumes the following form

$$\rho_{\lambda, \gamma}(\beta) = \begin{cases} \beta\lambda - \frac{\beta^2}{2\gamma} & \text{if } \beta \leq \gamma\lambda, \\ \frac{1}{2}\gamma\lambda^2 & \text{if } \beta > \gamma\lambda, \end{cases}$$

with $\gamma > 0$.

4   Lead Molecule Optimization

In this section we present the results achieved by Clu-EDO in designing and optimizing lead molecules. We build the procedure using a data set presented and analysed in [4,5]. These data have been used by the authors as a test environment to assess the effectiveness and the efficiency of new designs for lead optimisation. The data concern a library of 2500 molecules, identified by their chemical compositions (reagents) and experimental response (activity). These data represent the whole experimental space. Unlike the original dataset, the data we are dealing with are described by 22272 binary variables representing the presence or the absence of a particular molecular fragment. The response variable measures the biological activity of the reaction product. The aim of the analysis is to find the reaction whose product maximises the molecular activity. Results are analysed and compared to Lasso-EDO, the evolutionary design for optimization procedure based on the classical lasso procedure. Lasso-EDO belongs to the class of EDO-designs where the statistical model used for predictions is the lasso regression proposed by [13]. In this regression model the predictor variables are not grouped in clusters and the approach selects the most important predictors affecting the response variable without taking advantages of the grouping structure.

4.1   Estimation of Clusters and Penalized Regression Model

A preliminary analysis has shown linear dependencies among the predictors (fragments) of the dataset. A first reduction of dimensionality has been therefore achieved by detecting sets of linear dependent fragments and by considering just one representative fragment for each set. With this first analysis, we reduce the total amount of fragments to 4059. It is worth noting that in this research the dimension of the system p is 4059 while the number of observations n is at most 140. We therefore build Clu-EDO by firstly clustering the 4059 fragments into non-overlapping groups and then using the achieved cluster structure in the regression analysis.

The Bayesian non-parametric approach provides a solution with 85 clusters with a number of fragments inside each cluster varying between 1 and 232. The procedure is able to identify 2 singletons (i.e. clusters composed by only one fragment) as well as 6 big clusters containing more than one hundred of fragments each. The box-plot of the number of fragments in each cluster, i.e. cluster dimensionality, is displayed in Fig. 1. We notice that the 6 big clusters, which can be regarded as outliers, are identified by stars in Fig. 1.

A representation useful to evaluate the relation of the cluster fragment composition in terms of the biological activity of molecules is presented in Fig. 2. On the x-axis are reported all the 2500 molecules and on the y-axis the cluster identifications (cluster id); from left to right, molecules are displayed according to decreasing levels of activity. If one focuses on a specific row, i.e. on a particular cluster, it is possible to obtain information about which molecules have fragments included in the cluster. The rationale of the plot is to detect if some

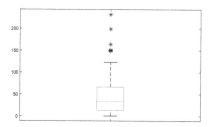

Fig. 1. The box-plot of the cluster dimensionality

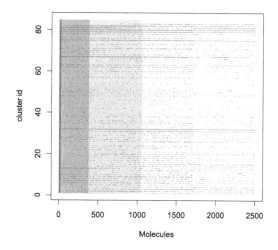

Fig. 2. Representation of the relation among fragments (blue dots), clusters (y-axis), and molecules (x-axis). Red, orange, yellow, beige and white regions identify molecules whose levels of activity are respectively 8 and in the intervals $[7.5; 8)$, $[6.5; 7.5)$, $[5; 6.5)$, and $[0; 5)$. (Color figure online)

clusters contain fragments that are mostly present in high activity level molecules. The size of the whole search space, i.e. 2500 molecules, makes it difficult to gather information from this plot. In Fig. 3 we focus only on the 24 (out of 2500) high-activity-level molecules, i.e. those molecules whose levels of activity are greater than 7.5. From this representation, it can be inferred that some clusters are empty or almost empty, revealing that the fragments belonging to those clusters do not characterize molecules with high activity. The 85 cluster structure is then used in the penalized regression analysis for selecting the most informative predictors related to the biological activity of the molecules.

4.2   Optimisation Results

We develop Clu-EDO by embedding in EDO procedure the statistical information obtained in the previous clustering and modelling steps. Each population

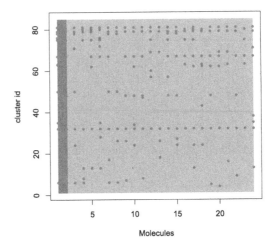

Fig. 3. Representation of the relation among fragments (blue dots), clusters (y-axis), and high-activity-level molecules (x-axis). Red and orange regions identify molecules whose levels of activity are 8 and in the interval [7.5; 8), respectively. (Color figure online)

of the evolutionary procedure is composed by 20 experimental points evolving across 7 generations, so that a total number of 140 molecules have been tested. In order to study the robustness of Clu-EDO with respect to changes in the initial population, we perform 1000 Monte Carlo simulations and we compare the results with Lasso-EDO built from the same initial populations. From Table 1 and the left panel of Fig. 4 we can notice that Clu-EDO is able to find the global optimum in 75% of the simulations, whereas the Lasso-EDO just in the 58% of the simulations. Similar results are also obtained by estimating the distribution of the best results with a non-parametric technique based on kernel estimators; see the right panel of Fig. 4. This figure shows that Clu-EDO (blue line) has a higher concentration of values around the 8 value. The comparison in performance between the two methods can also be made by considering a region of optimality (biological activity values greater than 7.5) instead of the single optimal value. Both approaches exhibit good performances as shown in Table 1. Comparisons between the two methods can also be made with respect to the evolution across generations. The evolution of the distribution of the activity values achieved in each generation is shown in Fig. 5. It should be pointed out that the global optimum value of the experimentation is indicated by the red dashed line. From this representation, we can notice that for Clu-EDO the response value is increasing in each generation and it is very close to the optimal solution. Moreover, the quartiles of the distribution improve generation after generation whereas for Lasso-EDO the median of the distribution of the best response value maintains the same value after the third generation. It is noteworthy that both approaches reach the optimal solution in almost one run for all the seven generations. These comparisons exhibit the very good performance of Clu-EDO in

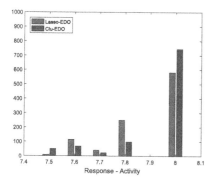
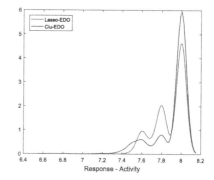

Fig. 4. Left panel: the histogram of the best activity values achieved in 1000 runs by the Lasso-EDO (red) and the Clu-EDO (blue). Right panel: the kernel estimate of the distribution of the best activity values achieved in 1000 runs by the Lasso-EDO (red) and the Clu-EDO (blue). (Color figure online)

Table 1. The number of runs out of 1000 in which the high-level activity values are achieved.

Design	Activity ≥ 7.5	Activity $= 8$
Lasso-EDO	995	582
Clu-EDO	981	745

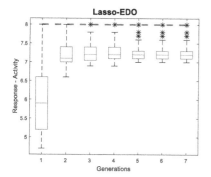
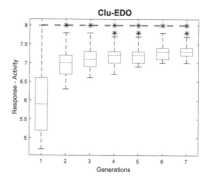

Fig. 5. Box-plot of the biological activity values through generations achieved in 1000 runs by the Lasso-EDO (left) and the Clu-EDO (right).

reducing the dimensionality of the system without any loss of information in fragment representation. Moreover, this reduction improves the capacity of the model to provide good predictive results and guide the evolution toward the optimum of the experimentation.

5   Concluding Remarks

The complexity in real phenomena, such as in drug discovery process, introduces a rise of interest in the research of new methodologies to analyse high-dimensional data characterized by a complex structure. In this paper we proposed a novel procedure to design experiments and modelling complex systems which integrates three strategies. The first strategy is the construction of clusters of predictors under the rationale that they have the same function or form similar patterns. The second strategy entails the selection of important predictors by exploiting the information obtained in the clustering phase. The last strategy is based on the evolutionary principles and the insight gained from the previous steps. Our findings suggest that combining evolutionary principles with the integration of a cluster structure in the statistical models is a promising approach to handle high-dimensional systems. We point out that the method could be easily adapted to handle experiments with a large number of continuous predictors. Further works will be devoted to investigate and compare other clustering alternatives.

Acknowledgements. The authors would like to acknowledge Professor Philip J. Brown and the GlaxoSmithKline Medicines Research Centre (UK) for the very fruitful collaboration in developing this research.

References

1. Fan, J., Lv, J.: A selective overview of variable selection in high dimensional feature space. Stat. Sin. **20**, 101–148 (2010)
2. Ma, Y., Zhu, L.: A review on dimension reduction. Int. Stat. Rev. **81**(1), 134–150 (2013)
3. Baragona, R., Battaglia, F., Poli, I.: Evolutionary Statistical Procedures: An Evolutionary Computation Approach to Statistical Procedures Designs and Applications. Springer, Heidelberg (2013)
4. Borrotti, M., De March, D., Slanzi, D., Poli, I.: Designing lead optimization of MMP-12 inhibitors. Comput. Math. Methods Med. **2014**, 1–8 (2014)
5. Pickett, S.D., Green, D.V.S., Hunt, D.L., Pardoe, D.A., Hughes, I.: Automated lead optimization of MMP-12 inhibitors using a genetic algorithm. ACS Med. Chem. Lett. **2**(1), 28–33 (2011)
6. Slanzi, D., De Lucrezia, D., Poli, I.: Querying Bayesian networks to design experiments with application to 1AGY serine esterase protein engineering. Chemometr. Intell. Lab. **149**, 28–38 (2015)
7. Santra, T.: A Bayesian non-parametric method for clustering high-dimensional binary data (2016). https://arxiv.org/pdf/1603.02494
8. Breheny, P., Huang, J.: Penalized methods for bi-level variable selection. Stat. Interface **2**(3), 369–380 (2009)
9. Lameijer, E.-W., Bäck, T., Kok, J.N., Ijzerman, A.D.P.: Evolutionary algorithms in drug design. Nat. Comput. **4**(3), 177–243 (2005)
10. Huang, J., Breheny, P., Ma, S.: A selective review of group selection in high-dimensional models. Stat. Sci. **27**(4), 481–499 (2012)

11. Liu, J., Wang, F., Gao, X., Zhang, H., Wan, X., Yang, C.: A penalized regression approach for integrative analysis in genome-wide association studies. J. Biom. Biostat. **6**(228), 1–7 (2015)
12. Ogutu, J.O., Piepho, H.P.: Regularized group regression methods for genomic prediction: Bridge, MCP, SCAD, group bridge, group lasso, sparse group lasso, group MCP and group SCAD. BMC Proc. **8**(Suppl. 5), S7 (2014)
13. Tibshirani, R.: Regression shrinkage and selection via the lasso. J. R. Stat. Soc. Series B Stat. Methodol. **58**(1), 267–288 (1996)
14. Zhang, C.-H.: Nearly unbiased variable selection under minimax concave penalty. Ann. Stat. **38**(2), 894–942 (2010)
15. Bühlmann, P., Rütimann, P., van de Geer, S., Zhang, C.H.: Correlated variables in regression: clustering and sparse estimation. J. Stat. Plan. Infer. **143**(11), 1835–1858 (2013)

Constraint-Based Modeling and Simulation
of Cell Populations

Marzia Di Filippo[1,4], Chiara Damiani[1,2(✉)], Riccardo Colombo[1,2],
Dario Pescini[1,3(✉)], and Giancarlo Mauri[1,2]

[1] SYSBIO Centre of Systems Biology, Piazza della Scienza 2, 20126 Milano, Italy
{chiara.damiani,dario.pescini}@unimib.it
[2] Dipartimento di Informatica, Sistemistica e Comunicazione,
Università degli Studi di Milano-Bicocca, Viale Sarca 336, 20126 Milan, Italy
[3] Dipartimento di Statistica e Metodi Quantitativi, Università degli Studi
di Milano-Bicocca, Via Bicocca degli Arcimboldi 8, 20126 Milan, Italy
[4] Dipartimento di Biotecnologie e Bioscienze, Università degli Studi
di Milano-Bicocca, Piazza della Scienza 2, 20126 Milan, Italy

Abstract. The intratumor heterogeneity has been recognized to characterize cancer cells impairing the efficacy of cancer treatments. We here propose an extension of constraint-based modeling approach in order to simulate metabolism of cell populations with the aim to provide a more complete characterization of these systems, especially focusing on the relationships among their components. We tested our methodology by using a toy-model and taking into account the main metabolic pathways involved in cancer metabolic rewiring. This toy-model is used as "individual" to construct a "population model" characterized by multiple interacting individuals, all having the same topology and stoichiometry, and sharing the same nutrients supply. We observed that, in our population, cancer cells cooperate with each other to reach a common objective, but without necessarily having the same metabolic traits. We also noticed that the heterogeneity emerging from the population model is due to the mismatch between the objective of the individual members and the objective of the entire population.

1 Introduction

A reprogramming of cellular energy metabolism has recently been included within the hallmarks [10] of cancer. An overall rewiring of metabolism is indeed fundamental to most effectively support the uncontrolled and enhanced growth characterizing all tumor cells.

An attention on the single molecules that are responsible for cancer onset fails to handle the non-linearity and complexity of cancer metabolic rewiring [3]. For this reason, metabolomics aims at concurrently identifying and quantifying the full set of metabolites that are present within a given cell or tissue type at a given time, thus providing a snapshot of the cell phenotype [4,11].

M. Di Filippo and C. Damiani—Equal contributors.

© Springer International Publishing AG 2017
F. Rossi et al. (Eds.): WIVACE 2016, CCIS 708, pp. 126–137, 2017.
DOI: 10.1007/978-3-319-57711-1_11

As a matter of fact, information and knowledge can be extracted from these large collections of data only by rationalizing and integrating them into computational predictive models. In this regard, constraint-based modeling has been by far the most applied technique to the study of metabolism. It indeed represents the best compromise between the purely qualitative information provided by graph-theory based topological models and the mechanistic details provided by kinetic modeling, which is currently impracticable for networks on a genome-scale. In particular, Flux Balance Analysis (FBA) – which exploits Linear Programming to identify the distribution of the metabolic flux that optimizes a metabolic objective – has extensively been applied to cancer research, as maximization of growth rate may accurately describe the objective driving cancer evolution [1,2,8]. Classic FBA is limited to the simulation of a single (or average) cell that is representative of the metabolism of the entire population this cell belongs to. This is a major drawback if we consider that a cell population is not necessarily homogeneous and various metabolic phenotypes may be generated. In fact, the heterogeneity characterizing cancer disease is not limited to the one existing among individual tumor types, but multiple sources of intratumor heterogeneity leading phenotypic differences among cells belonging to the same population exist. Unfortunately, many anti-cancer treatments are not able to deal with intratumor heterogeneity, drastically reducing their efficacy [9,14]. As a consequence, single-cell metabolomics techniques are currently under development as a promising strategy to unravel metabolic heterogeneity among cells belonging to the same tumor, which metabolomics hides as a result of average measurements of population behavior, by investigating singularly the role of distinct cell types within a given population. However, these kind of experiments are still at an early stage and numerous technical limitations remain to be solved [20].

To address the issue, we propose here an extension of constraint-based modeling to study metabolism of cell populations in order to provide a more complete characterization of these systems and to investigate relationships among their components. We assume that the heterogeneity emerging from a given cell population is due to the fact that the objectives of the individual members do not correspond to the objective of the entire population. As a proof of principle, we test our methodology with a toy-model of cancer metabolism that has been reconstructed based on the current knowledge on the metabolic pathways most involved in cancer metabolic rewiring.

2   Flux Balance Analysis and Flux Variability Analysis

Flux Balance Analysis allows to calculate the optimal flux distribution, which is the rate at which each of the R reactions of a network occurs at steady state.

By relying on a steady state assumption, according to which concentration of each of the M metabolites belonging to the network remains constant over time, FBA does not require any knowledge on enzymatic kinetic or metabolite concentrations. The application of further constraints on the system is used to reduce the number of putative flux distributions defining an allowable solution space in which any flux distribution may be equally acquired by the network. The optimization (maximization or minimization) of a specific objective function (e.g. maximization of adenosine triphosphate (ATP) or biomass production, minimization of nutrients utilization) allows to narrow the set of feasible solutions and to identify a single optimal flux distribution.

Given a $M \times R$ stoichiometric matrix S, whose element $s_{i,j}$ takes value $-\alpha_{ji}$ if the species S_i is a reactant of reaction j, $+\alpha_{ji}$ if species S_i is a product of reaction R_j and 0 otherwise - where $-\alpha_{ji}$ is the stoichiometric coefficient of reactant/product i in reaction j - the problem is postulated as a general Linear Programming formulation:

$$\text{maximize or minimize } Z = \sum_{i=1}^{R} w_i v_i \tag{1}$$

$$\text{subject to } S\boldsymbol{v} = \boldsymbol{0}, \ \boldsymbol{v}_{min} \leq \boldsymbol{v} \leq \boldsymbol{v}_{max}.$$

where w_i is the objective coefficient for flux v_i in vector $\boldsymbol{v}$; whereas the vectors $\boldsymbol{v}_{min}$ and $\boldsymbol{v}_{max}$ specify, respectively, the lower and upper boundaries of the admitted interval of each flux v_i. A negative value v_i conventionally indicates flux trough the backward reaction. To achieve mass balance in an open system, exchange of a given nutrient A with the environment is defined by unbalanced reactions in the form of $A <=> \emptyset$. For a more comprehensive description of FBA, the reader is referred to [15].

Frequently, although FBA only returns a single flux distribution, the constraints imposed on the system under investigation do not allow to obtain a unique solution, but may confine the solution space to a feasible set of alternative optimal flux distributions in which the same optimal flux value of the objective function may be reached through different but equally possible ways. In this context, Flux Variability Analysis (FVA) [12] returns the range of flux variability of each reaction, i.e. the allowable minimum and the maximum fluxes by each reaction, but it does not identify all the alternative optimal solutions. This task can be performed by exploiting recursive mixed integer linear programming (MILP) optimization, as proposed in [16].

3   A Proposal for Using the Constraint-Based Approach to Model Cell Populations

Metabolic networks reconstructed starting from genome annotation are today available for different organism, spanning from micro-organisms [7] to human metabolism. These networks may encompass virtually all reactions that can be catalyzed by the enzymes encoded by a given genome, or only a portion of them [17]. In order to fill the existing gap between the understanding of single cells function (represented by a single metabolic network) within a given tissue and their role when they are interacting with each other within a population, we propose to replicate N copies of the reference metabolic network with univocal names for metabolites and reactions, so to obtain a $(M \cdot N) \times (R \cdot N)$ stoichiometric matrix. For the exchange of intracellular nutrients with the environment (the extracellular matrix) of each of the N networks, the unbalanced reactions $A_i \Longleftrightarrow \emptyset$ are replaced by reactions in the form $A_i \Longleftrightarrow A_{medium}$ where A_i refers to metabolite A in network i, whereas A_{medium} refers to metabolite A in the extracellular matrix, to mimic the fact that cells in the population share the same resources. To achieve mass balance of the population model as an open system, a set of E exchange unbalanced reactions for metabolites within the extracellular matrix must be included. Note that the set of metabolites that the cells exchange with the extracellular matrix and the set of metabolites that the cell population share with the external environment do not necessarily coincide. A schematic representation of the population model is provided in Fig. 1. Once the $(M \cdot N) \times (R \cdot N + E)$ stoichiometric matrix and the vector of objective coefficients are obtained for the population model, standard FBA can be applied to obtain the distribution of flux across the N cells that maximize the population objective. For this purpose, a Python algorithm was implemented to automatically replicate a number of times any constraint-based model and generate the above defined population model. The resulting model is then exported to the Systems Biology Markup Language (SBML) format in order to be made suitable for simulation by any software that allows to perform linear programming optimization (e.g., COBRApy package [6], COBRA Toolbox [18]).

4   Results

As a proof of concept of our methodology, we constructed a generic and non-compartmentalized toy-model, that we refer to as "single entity core model", based on the current knowledge on the metabolic pathways most involved in cancer metabolic rewiring. This model consists of 45 reactions and 40 metabolites and includes the following metabolic pathways: glycolysis, production and consumption of lactate, tricarboxylic acid cycle (TCA cycle), oxidative phosphorylation (OXPHOS), pentose phosphate pathway (PPP), palmitate synthesis and beta-oxidation of fatty acids. Uptake reactions for the nutrients glucose and oxygen have been added as constraint to the model for defining the cell medium composition, and the maximization of the ATP yield has been chosen as objective function of the system, as we are focused on the reprogramming of energy metabolism of cancer cells.

We used this toy-model as building block for constructing the "population model" characterized by the interaction among individual components, all having the same stoichiometry and sharing the same glucose and oxygen supply. As for the single entity model, we chose the maximization of the overall ATP production as objective function of the whole system. We therefore investigated the potentialities of the constraint-based approach in the simulation of both the single entity and the population models in order to understand if this approach is able to highlight some differences between the two models in terms of their resulting flux distributions. The two models under investigation have the same

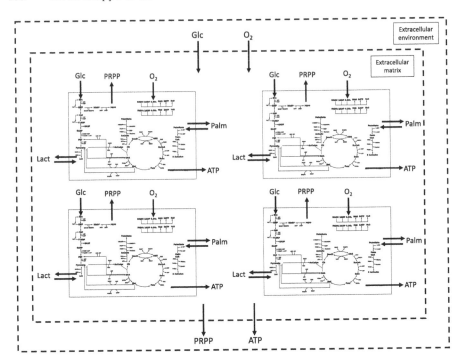

Fig. 1. Schematic representation of the population model. A single entity model is used as building block and replicated N times to obtain a network of metabolic networks. All the members belonging to the population model have the same topology and stoichiometry, share the same nutrients (in our case, glucose and oxygen) supply, and have the same reactions to exchange some metabolites with other components of the population (within a compartment referred to as "Extracellular matrix"), or with the external environment (referred to as "Extracellular environment"). Abbreviations: Glc, glucose; Lact, lactate; PRPP, phosphoribosyl pyrophosphate; Palm, palmitate; ATP, adenosine triphoshate, O_2, oxygen.

objective function, equal exchange (sink and demand) reactions and the same boundaries on nutrients uptake.

We applied FBA to obtain the ATP production yield – computed as the ratio between the objective function flux value and the number of entities included in the model – as a function of the simulated number of entities, including the classic case of one single entity. We observed that the objective function flux value of the population model increases proportionally with the increase in the number of its components (Fig. 2). The computed yield is, therefore, constant and is not affected by the number of entities. This outcome confirms that FBA on individual metabolic networks well approximates the average cell of an optimal population. In fact, the net flux distribution of the different cells perfectly mirrors the flux distribution obtained as a solution of the single FBA model (Fig. 3, panel A). However, the population model allows to investigate the tumor

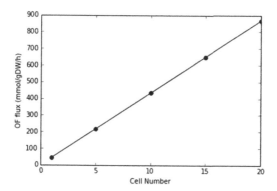

Fig. 2. Variation of objective function flux value of the population model according to the increasing in the number of its components. The plot shows how the objective function flux value of the population model ("OF flux") increases proportionally with the increase in the number of its members ("Cell Number").

population at a different level, elucidating the ways in which the average behavior can be achieved, how the individual cells may differ in their metabolism, and how different subpopulations of cells may interact with each other to attain the common goal.

4.1 Metabolic Heterogeneity Within Population Models

We shifted the focus toward a more in-depth study of how the flux distribution identified in the single entity model distributes among multiple cells within the population model. We wanted to understand whether FBA approach could highlight the heterogeneity factor that we know to be a long-established knowledge of cancer populations, or, in alternative, if the different components belonging to the system just share out the common good. In other words, we tested if a cooperative behavior could arise within cancer population or if all tumor cells behave the same way for achieving the common goal, which is an enhanced and uncontrolled growth and proliferation. In this regard, we used the toy-model to generate a population model consisting of 10 components, which are assumed to be single cells that are representative of the metabolism of distinct subpopulations this cells belong to, all having the same topology and stoichiometry, and sharing the same glucose and oxygen supply. We performed FBA simulations on this system maximizing its overall ATP production and then we exploited FVA analysis in order to explore the variability range of the system across the alternative ways for obtaining the same objective function flux value.

Given the same maximum amount of glucose and oxygen to the system, the reached steady state is characterized by a particular ratio between glucose and oxygen uptake flux value of 1:6, which is known to be the correct ratio so that one glucose molecule is completely oxidize by oxygen. We observed that glucose uptake flux value is adjusted based on the quantity of oxygen that is available in

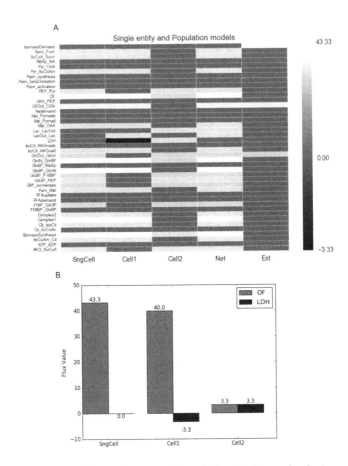

Fig. 3. Results obtained from the execution of Flux Balance Analysis on the single entity model and on the population model, by giving in both cases the same maximum amount of nutrients (glucose and oxygen). Panel A, Heatmap showing the flux values for the reactions of both the single entity and the population models. The column SngCell contains the flux values of the internal reactions belonging to the single entity model, the column Cell1 contains the flux values of the internal reactions of the first identified subpopulation of the population model, the column Cell2 contains the flux values of the internal reactions of the second identified subpopulation of the population model, the column Net contains the net flux values of the internal reactions of the two different subpopulations, whereas the column Ext contains the flux values for the exchange reactions of both the models. The color of each cell is proportional to the flux value of the corresponding reaction according to the gray chromatic scale on the right-hand side of each heatmap. Panel B, Bar plot showing the flux values for the objective function (referred to as "OF") and lactate production (referred to as "LDH") reactions in both the single entity and population models. Columns "Cell1" and "Cell2" refer to the two subpopulations of the population model.

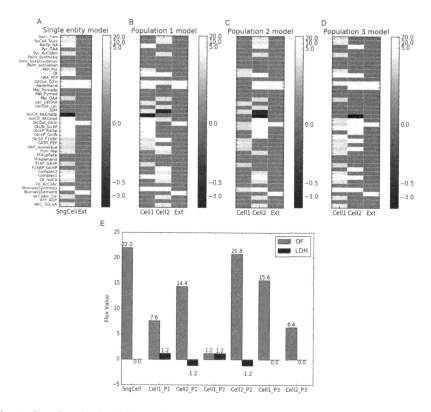

Fig. 4. Results obtained from the execution of Flux Balance Analysis on the single entity model and on the population model after a perturbation of the glucose/oxygen ratio. In all the heatmaps the color of each cell is proportional to the flux value of the corresponding reaction according to the gray chromatic scale on the right-hand side of each heatmap. Panel A, Heatmap showing the flux values for the reactions of the single entity model. The column SngCell contains the flux values of the internal reactions, whereas the column Ext contains the flux values of the exchange reactions. Panels B-C-D, Heatmaps showing the three alternative and equally optimal flux distributions identified in the population model. The column Cell1 contains the flux values of the internal reactions of the first identified subpopulation, the column Cell2 contains the flux values of the internal reactions of the second identified subpopulation, whereas the column Ext contains the flux values for the exchange reactions of the model. Panel E, Bar plot showing the flux values for the objective function (referred to as "OF") and lactate production (referred to as "LDH") reactions in both the single entity and population models. The graph shows that following the maximization of ATP production, a heterogeneity at objective function flux value level emerged in the population model between the two subpopulations of each of the three identified alternative optimal populations (columns "Cell1_P1" and "Cell2_P1", columns "Cell1_P2" and "Cell2_P2", columns "Cell1_P3" and "Cell2_P3"). The bar plot also shows, in all cases, that between the two interacting subpopulations of each population, the one that is responsible for the secretion of lactate in the medium produces ATP at a lower rate compared to the subpopulation in which lactate is consumed.

the medium, and that all the entities constituting the population under investigation seek to maximize the common good for satisfying the common aim. This aspect, showed by the analysis of the flux distribution of the population model, emerged together with the observation that maximization of the ATP production by the population model is obtained following the interaction between two distinct subpopulations which show a very different ATP production rate and differ in their energy generating pathways (Fig. 3). The first subpopulation, which corresponds to the hypoxic cancer cells, is composed by glucose-dependent cells that convert glucose into lactate that is then secreted in the medium; the second subpopulation, which corresponds to the better-oxygenated cancer cells, imports the lactate produced by the first subpopulation by using it as energy source instead of glucose, and is characterized by an active TCA cycle and respiratory chain. The flux distribution analysis showed that these two subpopulations do not contribute in an independent manner to the achievement of the common goal, but they cooperate with each other deriving mutual benefit from this interaction.

With changing environmental conditions as in Fig. 4, by perturbing the glucose to oxygen ratio and forcing the system towards more tumoral environmental conditions (i.e. constraining glucose uptake reaction flux to a higher level than that we found previously), the system reached different steady states having in common the fact that an increasing glucose uptake corresponds to a lowering of the objective function value. This happens because both there is not enough oxygen to completely oxidize glucose, and the individual can produce lactate whereas the entire population is not able. In addition to this result, we constantly noticed that, among the interacting subpopulations within the system, those that are responsible for the secretion of lactate in the medium, also produce ATP at a lower rate compared to the subpopulations in which lactate is consumed (Fig. 4, panel E). The analysis of flux variability, through FVA, showed that there is not just one single possible way by which different components belonging to the population model can interact with each other. On the contrary, for the purpose of maximizing the chosen objective function, three different scenarios (Fig. 4, panels B, C' and D), which represent alternative optimal solutions, emerged. This outcome strengthens the importance of the heterogeneity factor within cancer populations as a strategy developed for evolutionary reasons in order to resist to anti-tumor treatments.

5   Conclusions

To investigate heterogeneity within cellular populations and as a complementary approach to either single cell or standard metabolomics, we investigated the potentialities of a population model that is characterized by multiple interacting components, all having the same topology and stoichiometry, and sharing the same nutrients supply. These two elements were necessary for developing a methodology that allows to identify within a population model which are the best strategies able to promote cancer population growth and how many distinct subpopulations, characterized by different types of metabolism, interact with each other within the same tumor population. The advantage of performing FBA simulations on a population model compared to that on single entity model is the possibility of identifying distinct subpopulations having different phenotypes, but coexisting within the same system, and the possibility of better understanding the heterogeneity degree within a cancer population.

Through our approach, we came to the conclusion that the entire cancer population can be represented, at a first level, through a single entity model which provides a snapshot of the average behavior of the cell population, and at a second level, through a "network of metabolic networks", each of them representing the individual subpopulations. Indeed, just knowing the average behavior results in a limited outlook because the heterogeneity that might emerge within cancer population is not considered. Exploiting FBA method on a network consisting of multiple interacting components allowed us to observe that cancer cells cooperate with each other to reach a specific objective, and that they do not need to have the same metabolism type in order to reach the optimal value of objective function. Through our methodology we explored another level of complexity owned by cancer disease: the objective of the system does not correspond to the objectives of the individual entities since different subpopulations have different role within tumor tissue.

Since rewiring of energy-generating pathways and enhanced growth are closely related, the results here obtained following the ATP production maximization, may be generalized to the case of maximization of biomass production in cancer population. Accordingly, we can say that also the main metabolic trait that unifies all cancer cells (i.e., an uncontrolled and enhanced proliferation), is not the common objective for all individual cells belonging to the system. As stated by the cancer stem cell theory, the tumor growth is not driven by all cells belonging to the cancer population, but it is mainly sustained by only a specific portion of the tumor that consists in the so-called cancer stem cells [13,19].

Further analyses on more complex metabolic models will be performed to further validate our methodology and to investigate whether the observation that, among interacting subpopulations of a population model, those that are responsible for lactate consumption produce biomass at an higher level, holds even for more biologically grounded and comprehensive metabolic models. In principle, our modeling approach may be applied to genome-scale models. FBA is indeed computationally efficient even for very large networks, while the FVA computation can be sped up by parallel implementations. For example, the computation time to perform FVA on a model consisting of 2593 reactions and 2414 metabolites, by means of the COBRA Toolbox parallelized FVA function [18], is 27.34 s (Laptop Windows 10–64 bit Intel(R) Core i7-4710HQ CPU @ 2.50 GHz - RAM: 16.0 GB) and the computation time grows linearly with the model's size. Nevertheless, the actual problem of working with genome-scale models is the non-straightforward interpretation of the simulation outcomes, with particular regard to the typical large variability of optimal solutions, which may hinder the interpretation of the cooperation phenomena. Core metabolic models of specific aspects of metabolism may thus be more effective in uncovering system-level properties of cancer populations [5]. In conclusion, we would like to emphasize that the approach discussed here, is not tailored to just analyzing cancer cells populations, but it may be suitable for exploring, in general, how the interactions among more than one component (such as different types of healthy cells, bacteria, yeasts) may influence the overall behavior of a population for which a mismatch between the objective of the individual members and that of the entire population is assumed.

Acknowledgement. The institutional financial support to SYSBIO Center of Systems Biology - within the Italian Roadmap for ESFRI Research Infrastructures - is gratefully acknowledged. M.D. is supported by SYSBIO fellowship.

References

1. Agren, R., Bordel, S., Mardinoglu, A., Pornputtapong, N., Nookaew, I., Nielsen, J.: Reconstruction of genome-scale active metabolic networks for 69 human cell types and 16 cancer types using init. PLoS Comput. Biol. **8**(5), e1002518 (2012)
2. Agren, R., Mardinoglu, A., Asplund, A., Kampf, C., Uhlen, M., Nielsen, J.: Identification of anticancer drugs for hepatocellular carcinoma through personalized genome-scale metabolic modeling. Mol. Syst. Biol. **10**(3), 721 (2014)
3. Alberghina, L., Westerhoff, H.V.: Systems Biology: Definitions and Perspectives (Topics in Current Genetics), vol. 13. Springer, Heidelberg (2005)
4. Cazzaniga, P., Damiani, C., Besozzi, D., Colombo, R., Nobile, M.S., Gaglio, D., Pescini, D., Molinari, S., Mauri, G., Alberghina, L., et al.: Computational strategies for a system-level understanding of metabolism. Metabolites **4**(4), 1034–1087 (2014)
5. Di Filippo, M., Colombo, R., Damiani, C., Pescini, D., Gaglio, D., Vanoni, M., Alberghina, L., Mauri, G.: Zooming-in on cancer metabolic rewiring with tissue specific constraint-based models. Comput. Biol. Chem. **62**, 60–69 (2016)
6. Ebrahim, A., Lerman, J.A., Palsson, B.Ø., Hyduke, D.R.: COBRApy: Constraints-based reconstruction and analysis for python. BMC Syst. Biol. **7**(1), 74 (2013)
7. Feist, A.M., Herrgard, M.J., Thiele, I., Reed, J.L., Pallson, B.Ø.: Reconstruction of biochemical networks in microorganisms. Nat. Rev. Microbiol. **7**(2), 129–143 (2009)
8. Folger, O., Jerby, L., Frezza, C., Gottlieb, E., Ruppin, E., Shlomi, T.: Predicting selective drug targets in cancer through metabolic networks. Mol. Syst. Biol. **7**(1), 501 (2011)
9. Gerlinger, M., Rowan, A.J., Horswell, S., Larkin, J., Endesfelder, D., Gronroos, E., Martinez, P., Matthews, N., Stewart, A., Tarpey, P., et al.: Intratumor heterogeneity and branched evolution revealed by multiregion sequencing. New Eng. J. Med. **366**(10), 883–892 (2012)
10. Hanahan, D., Weinberg, R.A.: Hallmarks of cancer: the next generation. Cell **144**(5), 646–674 (2011)
11. Lee, J.M., Gianchandani, E.P., Papin, J.A.: Flux balance analysis in the era of metabolomics. Briefings Bioinform. **7**(2), 140–150 (2006)
12. Mahadevan, R., Schilling, C.: The effects of alternate optimal solutions in constraint-based genome-scale metabolic models. Metab. Eng. **5**(4), 264–276 (2003)
13. Marusyk, A., Polyak, K.: Tumor heterogeneity: causes and consequences. Biochimica et Biophysica Acta (BBA)-Rev. Cancer **1805**(117), 105–117 (2010)

14. Mohanty, A.K., Datta, A., Venkatraj, J.: Determining the relative prevalence of different subpopulations in heterogeneous cancer tissue. In: 2012 IEEE International Workshop on Genomic Signal Processing and Statistics (GENSIPS), pp. 95–96. IEEE (2012)

15. Orth, J.D., Thiele, I., Palsson, B.Ø.: What is flux balance analysis? Nat. Biotechnol. **28**(3), 245–248 (2010)

16. Reed, J.L., Palsson, B.Ø.: Genome-scale in silico models of e. coli have multiple equivalent phenotypic states: assessment of correlated reaction subsets that comprise network states. Genome Res. **14**(9), 1797–1805 (2004)

17. Ryu, J.Y., Kim, H.U., Lee, S.Y.: Reconstruction of genome-scale human metabolic models using omics data. Integr. Biol. **7**(8), 859–868 (2015)

18. Schellenberger, J., Que, R., Fleming, R.M.T., Thiele, I., Orth, J.D., Feist, A.M., Zielinski, D.C., Bordbar, A., Lewis, N.E., Rahmanian, S., et al.: Quantitative prediction of cellular metabolism with constraint-based models: the COBRA toolbox v2.0. Nat. Protoc. **6**(9), 1290–1307 (2011)

19. Yoo, M.H., Hatfield, D.L.: The cancer stem cell theory: is it correct? Mol. Cells **26**(5), 514 (2008)

20. Zenobi, R.: Single-cell metabolomics: analytical and biological perspectives. Science **342**(6163), 1243259 (2013)

Linking Alterations in Metabolic Fluxes with Shifts in Metabolite Levels by Means of Kinetic Modeling

Chiara Damiani[1,2($\boxtimes$)], Riccardo Colombo[1,2], Marzia Di Filippo[1,4],
Dario Pescini[1,3], and Giancarlo Mauri[1,2]

[1] SYSBIO Centre of Systems Biology, Piazza della Scienza 2, 20126 Milan, Italy
chiara.damiani@unimib.it
[2] Dipartimento di Informatica, Sistemistica e Comunicazione,
Università degli Studi di Milano-Bicocca, Viale Sarca 336, 20126 Milan, Italy
[3] Dipartimento di Statistica e Metodi Quantitativi,
Università degli Studi di Milano-Bicocca,
Via Bicocca degli Arcimboldi 8, 20126 Milan, Italy
[4] Dipartimento di Biotecnologie e Bioscienze,
Università degli Studi di Milano-Bicocca,
Piazza della Scienza 2, 20126 Milan, Italy

Abstract. The links between metabolic dysfunctions and various diseases or pathological conditions are being increasingly revealed. This revival of interest in cellular metabolism has pushed forward new experimental technologies enabling the characterization of metabolic phenotypes. Unfortunately, while large datasets are being collected, which encompass the concentration of many metabolites of a system under different conditions, these datasets remain largely obscure. In fact, in spite of the efforts to interpret alterations in metabolic concentrations, it is difficult to correctly ascribe them to the corresponding variations in metabolic fluxes (i.e. the rate of turnover of molecules through metabolic pathways) and thus to the up- or down-regulation of given pathways. As a first step towards a systematic procedure to connect alterations in metabolic fluxes with shifts in metabolites, we propose to exploit a Montecarlo approach to look for correlations between the variations in fluxes and in metabolites, observed when simulating the response of a metabolic network to a given perturbation. As a proof of principle, we investigate the dynamics of a simplified ODE model of yeast metabolism under different glucose abundances. We show that, although some linear correlations between shifts in metabolites and fluxes exist, those relationships are far from obvious. In particular, metabolite levels can show a low correlation with changes in the fluxes of the reactions that directly involve them, while exhibiting a strong connection with alterations in fluxes that are far apart in the network.

Keywords: Metabolic network modeling · Metabolic biomarkers · ODEs · Monte Carlo experiments · Metabolic fluxes prediction

© Springer International Publishing AG 2017
F. Rossi et al. (Eds.): WIVACE 2016, CCIS 708, pp. 138–148, 2017.
DOI: 10.1007/978-3-319-57711-1_12

1   Introduction

Metabolic profiling provides a readout of the biochemistry and physiological status of an individual or population, resulting from genetic factors and environmental exposure, that can be exploited in personalized medicine and public healthcare [8]. Analyzing the full complement of metabolites in body fluids such as urine and plasma using various spectroscopic methods allows indeed to link human metabolic variations (biomarkers) to disease risk factors. For instance, metabolite profiling has identified a key role for glycine in rapid cancer cell proliferation [9].

A major limit to the informative power of newly discovered metabolic biomarkers is posed by the impossibility to ascribe variations in metabolite concentrations to modifications in either their production or consumption pathways. Knowledge about the deregulation of the involved pathways, which might be effectively targeted for treatment of the disease, requires information about alterations in the metabolic fluxes (i.e. the rate at which a substance is transformed into another through a given reaction or pathway). Although isotopic labeling and metabolic flux analysis allow to indirectly derive such information through ad hoc laborious experiments, there is a quest for systematic and affordable high-throughput techniques.

Accordingly, metabolic network modeling is increasingly being exploited as a way to understand shifts in metabolism at a genome-wide level. As knowledge of kinetic parameters on a large scale is currently impracticable, constraint-based modeling is by far preferred to dynamic modeling [3]. The former models exploit a reasonable steady state assumption for internal metabolites and deal with fluxes only, while disregarding quantification of metabolites. Hence, the simulation outcomes can be hardly compared against the growingly rich availability of metabolomics data, with metabolic models mainly lacking validation. This downside pushes forward the urge for a strategy capable of linking variations in concentrations with variations in fluxes, and vice versa.

We propose to look for recurrent patterns or "rules" of association between the changes in fluxes and metabolites observed in the steady states possibly obtained when simulating the response to a given perturbation (e.g. change in a metabolite level) of a fixed metabolic network stoichiometry for a large ensemble of randomly generated kinetic parameters. The idea moves from the hypothesis that properties shared by many randomly parameterized models should be determined by the network stoichiometry and will thus pertain also to the real biological system. Stoichiometric core models that focus on the main carbon and nitrogen sources and on the metabolic events that sustain growth have proven able to closely reproduce the metabolic properties of genome-wide networks [7,12], while reducing network complexity, and may therefore well serve our purpose. As a proof of principle, we investigate the dynamics of a simplified model of yeast glucose metabolism, which was designed in [6] to take into account only the pathways mainly involved in the emergence of the Crabtree effect.

2   Methods

Monte Carlo Simulations

Given a metabolic network model, defined by as set $\mathcal{R} = \{r_1, r_2, \ldots, r_R\}$ of R reactions and as set $\mathcal{M} = \{x_1, x_2, \ldots, x_M\}$ of M metabolites, we randomly generated N sets of kinetic constants $\mathcal{K}_1 = \{k_1, k_2, \ldots, k_R\}$, $\mathcal{K}_2 = \{k_1, k_2, \ldots, k_R\}, \ldots, \mathcal{K}_N = \{k_1, k_2, \ldots, k_R\}$ for the model reaction rates.

As a first approximation, the elementary mass action law was assumed for every reaction rate. For each parameter set $\mathcal{K}_p$, with $p = 1, \ldots, N$, we performed an ODEs-based deterministic simulation of the model until the system reached a steady state.

The simulation was repeated under two different nutritional conditions, a and b, each corresponding to a different availability (imposed constant concentration value) of a given nutrient (namely, glucose).

For both conditions, we then calculated the flux values $v_{i,p}^a$ and $v_{i,p}^b$ for each $r_i \in \mathcal{R}$ at steady state, according to the mass action relation:

$$v_i = k_i \prod_{w=1}^{M} [x_w]^{\alpha_{wi}} \tag{1}$$

where k_i is the kinetic constant of the rate of reaction r_i, $[x_w]$ is the concentration of species w and α_{wi} the stoichiometric coefficient according to which species w participate in reaction r_i.

Experimental Setting

- The kinetic parameters were randomly generated from a uniform distribution in [0,1).
- Integration of ODEs was executed by using the LSODA solver [11].
- Based on the experimental evidence that *in vivo* metabolism reaches a steady state within few seconds [15], we run the simulation for a simulated time of 50 s. For the sake of efficiency, the time series of each metabolite concentration, integrated by means of an adaptive integration step method, was evenly sampled a thousand of times.
- Cellular volume was set to 1.66667e-15 *l* according to literature on *Saccharomyces cerevisiae* [4].
- The initial concentration of all metabolites in the network was mined from literature [4] and set according to the average values in Smallbone et al. [14], and Canelas et al. [2].
- Species corresponding to nutrients coming from the environment are simulated "in feed" (i.e. with a constant buffering). Nutrients considered: oxygen and glucose.
- To establish whether the system reached a steady state: we first calculated for every metabolite the standard deviation (σ) of the value of its concentration for the last 10% of its temporal evolution. We then evaluated the ratio of the

sum of the σs to the number of species not "in feed". If this ratio was less than 0.01 we considered the system at steady state and took the corresponding random parametrization into account, otherwise we disregarded it. We iterated the procedure until collecting $N = 10^5$ parametrization, thus ignoring about 23000 sets of random kinetic constants.
- For each random set of kinetic constants, we performed 10 different simulations evenly spanning the glucose interval $[0, 25]$ mMol.

From Constraints on Concentrations to Constraints on Fluxes

In order to seek possible relationships between the changes in fluxes and in metabolites observed as a response to a network perturbation (i.e. change in a metabolite level), for each parameter set $\mathcal{K}_p$, with $p = 1, \ldots, N$, we first computed the difference in the fluxes ($\delta v_{i,p} = v_{i,p}^a - v_{i,p}^b$, for each reaction $r_i \in \mathcal{R}$), and in the concentrations ($\delta x_{j,p} = [x_{j,p}^a] - [x_{j,p}^b]$, for each metabolite $x_j \in \mathcal{M}$). The differences were computed between any pair of steady states, corresponding to the same parameters set, but to different glucose availabilities, as obtained with the Monte Carlo approach described above.

Once the vectors $\delta v_i = \{\delta v_{i,1}, \delta v_{i,2}, \ldots, \delta v_{i,N}\}$ and $\delta x_j = \{\delta x_{j,1}, \delta x_{j,2}, \ldots, \delta x_{j,N}\}$ were obtained, we computed, as a first step, a Pearson product-moment correlation coefficient ρ (Eq. 2) between δv_i and δx_j for any pair i, j, across all the parameter sets $\mathcal{K}_p$, with $p = 1, \ldots, N$, limiting thus the analysis to linear relationships.

$$\rho_{\delta v_i, \delta x_j} = \frac{\text{cov}(\delta v_i, \delta x_j)}{\sigma_{\delta v_i} \sigma_{\delta x_j}} \tag{2}$$

To assess the robustness of the discovered relationships to the extent of the perturbation, for each pair of shifts, we evaluated their correlation for different magnitudes of the glucose variation.

The average correlation coefficient $\hat{\rho}$ of all pairs was also computed, as follows:

$$\hat{\rho} = \frac{1}{N} \sum_{i,j \in \mathcal{A}} |\rho_{\delta v_i, \delta x_j}| \tag{3}$$

where $\mathcal{A}$ is the set of all possible pairs i, j.

3   Results

The procedure described above was applied to the metabolic network of yeast glucose metabolism [6] depicted in Fig. 1.

The Pearson correlation coefficients between shifts in fluxes and shifts in metabolites – when depleting glucose from a baseline concentration of 25 mM to zero – are reported in the heatmap in Fig. 2. Notably, several pairs show a strong correlation, suggesting that, at steady state, alterations in metabolites are somehow constrained by alterations in fluxes, or vice versa. Intriguingly, covariations

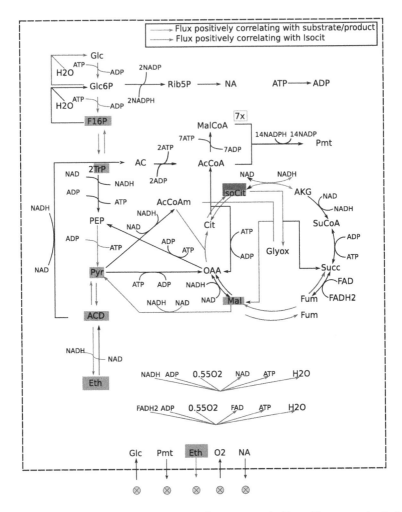

Fig. 1. Core metabolic network of yeast glucose metabolism (figure readapted from [4]). Directional arrows indicate reactions transforming substrate metabolite(s) into product metabolite(s). Coloring of arrows (when different from black) matches the color of the metabolite that significantly correlates with the reaction's flux. Abbreviations of metabolites: 2TrP = 2 Triose phosphate (Dihydroxyacetone phosphate + Glyceraldehyde 3-phosphate), AC = Acetate, ACD = Acetaldehyde, AcCoA = Acetyl-CoA, ADP = Adenosine diphosphate AKG = Alpha-ketoglutarate, ATP = Adenosine triphosphate Cit = Citrate, Eth = Ethanol, F16P = Fructose 1,6-bisphosphate, FAD = Flavin adenine dinucleotide, FADH2 = Flavin adenine dinucleotide (hydroquinone form), Fum = Fumarate, Glc = Glucose, Glc6P = Glucose 6-phosphate, H2O = Water, Isocit = Isocitrate, Mal = Malate, MalCoA = Mallonyl-CoA, NA = nucleic acids, NAD = Nicotinamide adenine dinucleotide (oxidized), NADH = Nicotinamide adenine dinucleotide (reduced), NADP = Nicotinamide adenine dinucleotide phosphate (oxidized), NADPH = Nicotinamide adenine dinucleotide phosphate (reduced), O2 = Oxygen, OAA = Oxaloacetate, PEP = Phosphoenolpyruvate, Pmt = palmitate, PYR = Pyruvate, Rib5P = Ribose 5-phosphate, Succ = Succinate, SuCoA = Succinyl-CoA. (Color figure online)

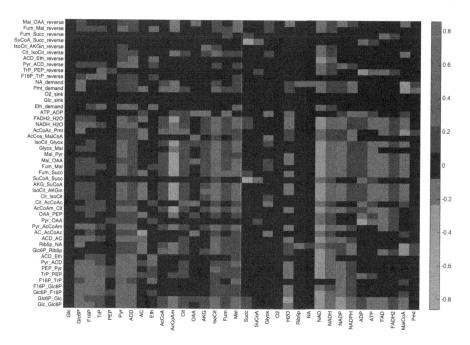

Fig. 2. Rows: fluxes; columns: metabolites; color: correlation coefficient. Metabolites are named as in Fig. 1. Reaction are indicated as substrate_product, with substrate and product being one among the reaction substrates and products respectively. Reverse reactions are considered separately and are indicated with the suffix '_reverse'. Insertion of a metabolite within the network is indicated by the suffix '_sink', whereas removal of a metabolite from the network is indicated with the suffix '_demand'. Note that NaN values, corresponding to fluxes with 0 variance (e.g. "in feed" nutrients), are replaced with 0.

may be not obvious at all: for instance the flux from Pyr and OAA to $AccoA$ shows a stronger positive correlation with a metabolite that is further in the network ($Isocit$) than with the reaction product; whereas, the flux from $SuCoA$ to $Succ$ surprisingly shows a negative correlation with the reaction product. We verified that, as expected given the considerable size (10^5) of the sample, all the computed correlation coefficients are statistically significant (p-value < 0.001) except for very weak correlations ($\rho \sim 0$).

Although different strategies might be followed to assess the robustness of the predicted correlations, we assessed the variation in the estimated correlation, when different magnitudes of the nutrient perturbations are simulated (from 25 to $\{0, 2.8, 16, 19.4, 22.2\}$ millimoles). We first analyzed how the distribution of the correlation coefficients obtained for all reactions is affected by the extent of the glucose perturbation. We clearly observed that the smaller the perturbation, the higher the average value of the correlation between shifts in metabolites and fluxes (Fig. 3A). Higher averages reflect the higher variability of the correlation strength between different pairs $\delta v_i, \delta x_j$ (boxplots in Fig. 3B). We did expect a low variability for the perturbation $\delta Glc = 25\,\mathrm{mM}$, as it corresponds to a total glucose depletion. As glucose is the unique carbon source in the network, flux throughout the entire network should be impossible for $Glc = 0\,\mathrm{mM}$, with the exception of futile cycles. The perturbation from 0 to $25\,\mathrm{mM}$ would therefore result in positive variations for all fluxes. Negative variations would be observed only for fluxes in the reverse reaction, which are treated as negative values. In fact, the number of positive correlations ($\rho > 0$) seems to decrease with the magnitude of the perturbation, accordingly the number of negative correlations ($\rho < 0$) apparently increases with the δGlc (Fig. 4).

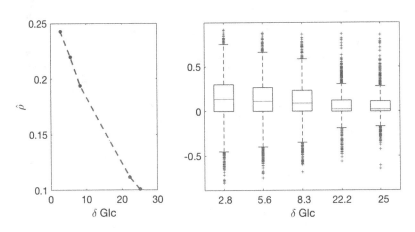

Fig. 3. (A) Average correlation coefficient as a function of the perturbation extent (δGlc). (B) Boxplot for the data in A.

When considering correlations above ± 0.5 as strong, we also observed that the stronger the perturbation the lower the number of strong correlations found (Fig. 4). This result is probably a consequence of the undetectable negative correlations for large perturbations. It is worth noticing that negative correlations occur way less frequently than positive ones, reasonably in view of the positive effect of carbon on total flux.

Table 1 ranks the reaction-metabolite pairs according to the absolute value of their average correlation. The average (indicated as $\hat{\rho}_{\delta v_i, \delta x_j}$) and standard deviations (st. dev.) of ρ across the 5 perturbations are indicated. The table also reports the number of perturbation magnitudes for which a strong correlation has been observed for the pair at issue (out of 5). We also verified that the correlation has always the same sign across the 5 values.

Several pairs show a strong correlations with a low dispersion. If we restrict our analysis to the pairs $\delta v_i, \delta x_j$ that exhibit a strong correlation ($\rho_{\delta v_i, \delta x_j} > \pm 0.5$) regardless of the perturbation extent ($N\,selected\,times \equiv 5$), we can meaningfully represent the relationship between metabolite and flux on the metabolic map in Fig. 1. It can be observed that, although some fluxes (red arrows) do

Table 1. List of metabolite-flux pairs with $\hat{\rho}_{\delta v_i,\delta x_j} \geq \pm 0.51$ ranked by $\mid \hat{\rho}_{\delta v_i,\delta x_j} \mid$

Rxn	Met	$\hat{\rho}_{\delta v_i,\delta x_j}$	St. dev.	N^o selected times	N^o positive times
Cit_IsoCit_reverse	IsoCit	0.86	0.01	5.00	5.00
Cit_IsoCit	IsoCit	0.84	0.05	5.00	5.00
Fum_Mal_reverse	Mal	0.82	0.01	5.00	5.00
F16P_TrP	TrP	0.76	0.04	5.00	5.00
Pyr_ACD	ACD	0.76	0.03	5.00	5.00
Fum_Mal	Mal	0.76	0.06	5.00	5.00
Pyr_ACD_reverse	ACD	0.75	0.01	5.00	5.00
Eth_EthOUT	Eth	0.74	0.17	5.00	5.00
F16P_TrP_reverse	TrP	0.73	0.05	5.00	5.00
IsoCit_Glyox	IsoCit	0.73	0.10	5.00	5.00
ACD_Eth	Eth	0.72	0.20	4.00	5.00
Pyr_ACD	Pyr	0.71	0.12	5.00	5.00
FADH2_H2O	IsoCit	0.71	0.12	5.00	5.00
Glyox_Mal	IsoCit	0.68	0.11	5.00	5.00
Fum_Succ	IsoCit	0.67	0.12	5.00	5.00
Mal_Pyr	IsoCit	0.67	0.09	5.00	5.00
AcCoAm_Cit	IsoCit	0.66	0.14	5.00	5.00
Glc6P_GlcIN	IsoCit	0.66	0.11	5.00	5.00
Fum_Mal	IsoCit	0.65	0.16	3.00	5.00
NADH_H2O	IsoCit	0.64	0.11	5.00	5.00
Mal_OAA	Mal	0.64	0.26	3.00	5.00
GlcIN_Glc6P	IsoCit	0.63	0.11	5.00	5.00
ACD_AC	Pyr	0.62	0.15	3.00	5.00
ACD_Eth_reverse	Eth	0.62	0.24	3.00	5.00
IsoCit_AKGin	IsoCit	0.61	0.12	4.00	5.00
Pyr_AcCoAm	IsoCit	0.61	0.10	4.00	5.00
PEP_Pyr	Pyr	0.60	0.14	3.00	5.00
AKG_SuCoA	IsoCit	0.60	0.12	3.00	5.00
TrP_PEP	Pyr	0.59	0.15	3.00	5.00
Pyr_ACD_reverse	Pyr	0.59	0.09	3.00	5.00
Mal_OAA	IsoCit	0.57	0.12	3.00	5.00
ATP_ADP	ATP	0.57	0.27	3.00	5.00
ATP_ADP	ADP	−0.57	0.27	3.00	0.00
Glc6P_F16P	F16P	0.56	0.03	5.00	5.00
NADH_H2O	Pyr	0.55	0.14	3.00	5.00
GlcIN_Glc6P	Pyr	0.55	0.14	3.00	5.00
Mal_OAA_reverse	Mal	0.53	0.23	3.00	5.00
FADH2_H2O	Mal	0.53	0.20	3.00	5.00
Cit_IsoCit	Mal	0.53	0.22	3.00	5.00
Glc6P_F16P	Pyr	0.52	0.14	3.00	5.00
AcCoAm_Cit	Mal	0.52	0.21	3.00	5.00
Fum_Succ	Mal	0.52	0.21	3.00	5.00
Glc6P_GlcIN	Pyr	0.52	0.14	3.00	5.00
PEP_Pyr	ACD	0.51	0.09	3.00	5.00
Pyr_OAA	Eth	0.51	0.32	3.00	5.00
NADH_H2O	NADH	0.51	0.10	3.00	5.00
NADH_H2O	NAD	−0.51	0.10	3.00	0.00
TrP_PEP	NADH	0.51	0.14	3.00	5.00
TrP_PEP	NAD	−0.51	0.14	3.00	0.00
GlcIN_Glc6P	NADH	0.51	0.10	3.00	5.00
GlcIN_Glc6P	NAD	−0.51	0.10	3.00	0.00
ACD_AC	ACD	0.51	0.08	3.00	5.00
F16P_Glc6P	F16P	0.51	0.01	3.00	5.00
PEP_Pyr	NADH	0.51	0.14	3.00	5.00
PEP_Pyr	NAD	−0.51	0.14	3.00	0.00
Glc6P_GlcIN	Mal	0.51	0.19	3.00	5.00

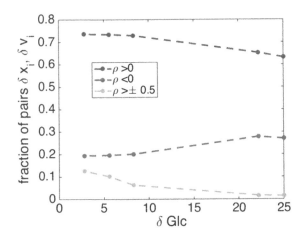

Fig. 4. Fraction of values of ρ, computed for all the metabolite-reaction pairs and across the 10^5 parametrizations, indicating a positive correlation ($\rho > 0$), a negative correlation ($\rho > 0$) or a strong positive/negative correlation $\rho > \pm 0.5$, as a function of δGlc.

directly correlate with the reaction substrate or product, the correlation is typically positive for both substrate and product. Although biologists may typically expect an increase in flux to be associated with a decrease in substrate, when thinking at the mass action rate law, it is not surprising that the chemical reaction is directly proportional to the product of concentrations of the reactants. Strikingly, all the selected fluxes that do not directly correlate with their substrate/product seem to correlate with the same metabolite: *IsoCit* (isocitrate). The central role of isocitrate is apparent in Fig. 1 and may be regarded as an emergent property of the network dynamics. This result implies that simplistic interpretations that ascribe the decrease/increase in the concentration of isocitrate to variations in fluxes that are directly responsible for its production/consumption should be avoided.

4   Conclusions and Perspectives

Systems biology typically exploits constraint-based modeling, and in particular Flux Balance Analysis, to investigate the properties of a metabolic network [10]. Stoichiometric genome-wide models of metabolism have found application especially in Health and Medicine (see for example [1]). A major benefit of this kind of models is that they do not require any knowledge about the kinetic parameters, while a major drawback is the impossibility to provide information on the transient evolution of the system and on the metabolic concentrations either at steady state or in the transient phase.

Although an attempt to cope with the incapability of FBA to provide information on metabolic concentrations has already been put forward [13], the proposed method provides predictions on extracellular metabolites only.

To cope with the absence of accurate knowledge on parameter values, Monte Carlo based strategies have been proposed, in which the kinetic parameters are repeatedly sampled from specific probability distributions and used for multiple parallel simulations, to predict the effects of perturbations on signaling networks [16], whose simulation cannot take advantage of the constraint-based approach.

We applied the Monte Carlo approach to simulate the dynamics of an ODEs core model of glucose yeast metabolism with the aim of uncovering the possible relationships between the shifts in metabolite concentrations and in flux values, observed as a response to a network perturbation. Specifically, we simulated a perturbation in glucose availability.

We identified some linear relationships between variations in fluxes and in metabolites. In spite of the extreme simplicity of the metabolic network – e.g. only one carbon source is considered and the contribution of nitrogen and carbon to protein synthesis is not taken into account – and of the linearity restriction, we found some unforeseen correlations. These results on the one hand emphasize the inefficacy of naive predictions of flux variations from metabolites alterations and viceversa, on the other hand they support the hypothesis that strong complex relationships exist which, if disentangled, may make such attempt less daring.

The proposed approach represents a small step forward in the achievement of this ambitious goal and more has yet to be done. The next step might involve the analysis of non linear relationships between shifts in fluxes and metabolites, by measuring the mutual dependence between two variables with measures rooted in information theory, as for instance the Mutual Information.

Although the total computational time to produce the data set was reasonable (5.5 h to run ODEs simulations on a MacBookPro with CPU 2.6 GHz Intel Core i7, RAM 16 GB and to produce 268 Mb of data), the computation of a large number of model trajectories may benefit form the exploitation of the GPU-accelerated algorithms proposed in [5].

The methodology will eventually further be developed to predict variations in fluxes, given an experimentally observed variation in metabolites, as per metabolic profiling experiments.

References

1. Agren, R., Mardinoglu, A., Asplund, A., Kampf, C., Uhlen, M., Nielsen, J.: Identification of anticancer drugs for hepatocellular carcinoma through personalized genome-scale metabolic modeling. Mol. Syst. Biol. **10**(3), 721 (2014)
2. Canelas, A.B., van Gulik, W.M., Heijnen, J.J.: Determination of the cytosolic free nad/nadh ratio in saccharomyces cerevisiae under steady-state and highly dynamic conditions. Biotechnol. Bioeng. **100**(4), 734–743 (2008)
3. Cazzaniga, P., Damiani, C., Besozzi, D., Colombo, R., Nobile, M.S., Gaglio, D., Pescini, D., Molinari, S., Mauri, G., Alberghina, L., et al.: Computational strategies for a system-level understanding of metabolism. Metabolites **4**(4), 1034–1087 (2014)
4. Colombo, R., Damiani, C., Mauri, G., Pescini, D.: Ensembles of parametrizations to investigate the crabtree phenotype by constraining mechanism-based simulations. In: Proceedings of CIBB 2016 (2016)

5. Cumbo, F., Nobile, M., Damiani, C., Colombo, R., Mauri, G., Cazzaniga, P.: Cosys: computational systems biology infrastructure. In: Proceedings of CIBB 2016 (2016)

6. Damiani, C., Pescini, D., Colombo, R., Molinari, S., Alberghina, L., Vanoni, M., Mauri, G.: An ensemble evolutionary constraint-based approach to understand the emergence of metabolic phenotypes. Nat. Comput. **13**(3), 321–331 (2014)

7. Di Filippo, M., Colombo, R., Damiani, C., Pescini, D., Gaglio, D., Vanoni, M., Alberghina, L., Mauri, G.: Zooming-in on cancer metabolic rewiring with tissue specific constraint-based models. Comput. Biol. Chem. **62**, 60–69 (2016)

8. Holmes, E., Wilson, I.D., Nicholson, J.K.: Metabolic phenotyping in health and disease. Cell **134**(5), 714–717 (2008)

9. Jain, M., Nilsson, R., Sharma, S., Madhusudhan, N., Kitami, T., Souza, A.L., Kafri, R., Kirschner, M.W., Clish, C.B., Mootha, V.K.: Metabolite profiling identifies a key role for glycine in rapid cancer cell proliferation. Science **336**(6084), 1040–1044 (2012)

10. OBrien, E.J., Monk, J.M., Palsson, B.O.: Using genome-scale models to predict biological capabilities. Cell **161**(5), 971–987 (2015)

11. Petzold, L.: Automatic selection of methods for solving stiff and nonstiff systems of ordinary differential equations. SIAM J. Sci. Stat. Comput. **4**(1), 136–148 (1983). http://dx.doi.org/10.1137/0904010

12. Resendis-Antonio, O., Checa, A., Encarnación, S.: Modeling core metabolism in cancer cells: surveying the topology underlying the warburg effect. PloS one **5**(8), e12383 (2010)

13. Shlomi, T., Cabili, M.N., Ruppin, E.: Predicting metabolic biomarkers of human inborn errors of metabolism. Mol. Syst. Biol. **5**(1), 263 (2009)

14. Smallbone, K., Messiha, H.L., Carroll, K.M., Winder, C.L., Malys, N., Dunn, W.B., Murabito, E., Swainston, N., Dada, J.O., Khan, F., et al.: A model of yeast glycolysis based on a consistent kinetic characterisation of all its enzymes. FEBS Lett. **587**(17), 2832–2841 (2013)

15. Theobald, U., Mailinger, W., Baltes, M., Rizzi, M., Reuss, M.: In vivo analysis of metabolic dynamics in saccharomyces cerevisiae: I. experimental observations. Biotechnol. Bioeng. **55**(2), 305–316 (1997)

16. Wierling, C., Kühn, A., Hache, H., Daskalaki, A., Maschke-Dutz, E., Peycheva, S., Li, J., Herwig, R., Lehrach, H.: Prediction in the face of uncertainty: a Monte Carlo-based approach for systems biology of cancer treatment. Mutat. Res. Genet. Toxicol. Environ. Mutagen. **746**(2), 163–170 (2012)

Systems Chemistry and Biology

A Strategy to Face Complexity: The Development of Chemical Artificial Intelligence

Pier Luigi Gentili[✉]

Department of Chemistry, Biology and Biotechnology, University of Perugia, Perugia, Italy
pierluigi.gentili@unipg.it

Abstract. Nowadays, science is spurred to win the Complexity Challenges. There are challenges regarding Natural Complexity. But there are also challenges regarding Computational Complexity. A strategy to face both of them consists in developing Chemical Artificial Intelligence. Its development requires an analysis of the Human Nervous System and Human Intelligence at three levels; at the (i) Computational, (ii) Algorithmic, and (iii) Implementation levels, respectively. The effectiveness of this approach is demonstrated by showing three ways for implementing Fuzzy logic at the molecular level.

1 Introduction

Science is striving to tackle the Complexity Challenges. There are two types of Complexity Challenges: one type regards Natural Complexity and the other Computational Complexity.

Natural Complexity is synonym of intricacy. In fact, a Natural Complex System is a network of nonlinear and often adaptive interactions among unique elements. Such nonlinear interactions operate at different spatial scales and give rise to hierarchical structures. They also operate at different time scales and give rise to either stationary or periodic or aperiodic dynamics. Their dynamics are extremely sensitive on the contour conditions. As Warren Weaver forebodingly alleged in 1948, the Complexity of a particular system is the degree of difficulty in predicting the properties of the system when the properties of system's parts are given [1]. A prototype of complex system is the planet earth with its atmosphere, its oceans, its biosphere, and its crust. The complexity grounds on the networks of gradients and flows that make the weather forecast really difficult in the medium term and even impossible in the long term. Other examples of complex systems are the living beings, both unicellular and multi-cellular. In this case, the complexity grounds on the networks of the huge number of biochemical reactions occurring within a cell. Among the living beings, the humans are the most complex, due to their intelligence and their power of computing with words and giving rise to social and economic organizations, which are other examples of Complex Systems. We need to understand if there are general principles governing the behavior of Natural Complex Systems because we still fail to predict most of the catastrophic events in our planet; we encounter great difficulties in protecting our environment and our ecosystems; we need to find innovative solutions for the world energy issues; we

© Springer International Publishing AG 2017
F. Rossi et al. (Eds.): WIVACE 2016, CCIS 708, pp. 151–160, 2017.
DOI: 10.1007/978-3-319-57711-1_13

need to discover new effective treatments for incurable diseases, and we need, also, to contrive new solutions for the problems of economic and social stability.

As far as the Computational Complexity is concerned, we know that there exist exponential problems that cannot be solved accurately and in reasonable time with our current computing facilities. Very large exponential problems are transformed in Non-Deterministic Polynomial (NP) problems, i.e. in non-deterministic recognition problems, feeling satisfied after finding acceptable but not necessarily exact solutions. Of course, a compelling challenge is to propose new algorithms and/or design new computing machines to solve accurately and in reasonable time even the exponential problems of large dimensions. Finally, there exists another Computational Complexity challenge that stays at the border with Natural Complexity. It is the challenge of formulating universally valid and effective algorithms for the recognition of variable patterns, like human faces and voices, handwritten numbers and cursive words, fingerprints, patterns in medical diagnosis, patterns in apparently uncorrelated experimental data.

How do we tackle the Complexity Challenges? Since when we face Complexity, we are usually supposed to handle a huge number of data, we need to speed up our computational rate and/or find out new ways to process data. At the moment, there are two strategies to be followed. One strategy consists in improving current electronic computers. The other consists in developing the interdisciplinary research line of Natural Computing [2]. Researchers working on Natural Computing draw inspiration from nature to propose new algorithms, new materials to compute and new models to interpret Complexity, based on the rationale that every natural transformation is a kind of computation.

We are contributing to Natural Computing by focusing our attention on the human nervous system that has human intelligence as its own emergent property. The human nervous system is a complex network of billions of nerve cells that allows us to handle both accurate and vague information (by computing not only with numbers but also with words); to make decisions in complex situations (when we encounter many intertwined variables); to recognize quite easily variable patterns. Therefore, it is clear that to tackle the Complexity Challenges, it is worthwhile trying to deeply understand the working principles of human intelligence in order to reproduce them, artificially. To mimic the performances of human intelligence, we are using not electronic circuits and software, but chemicals and chemical reactions, i.e. wetware. In other words, we are developing Chemical Artificial Intelligence [3].

2 Methodology

To succeed in our project of developing Chemical Artificial Intelligence, we are following a methodology that has been proposed by the cognitive scientists Gallistel and King [4] and by the neuroscientist Marr [5] as an effective methodology to deal with any complex systems. It requires an analysis of a complex system like the human nervous system, at three levels. First, an analysis at the computational level. Such analysis consists in determining the inputs, the outputs and the computations that the system performs and its logic. Then, an analysis at the algorithmic level follows. It consists in formulating algorithms that might carry out those computations. Finally, an analysis at the implementation level starts. It consists in looking for mechanisms of the kind that would make the algorithms work.

In this contribution, we are presenting some results of the analysis of the human nervous system at the computational and algorithmic levels, highlighting structural and functional analogies that it shares with Fuzzy logic. Finally, three strategies to implement Fuzzy logic by molecules will be explained.

3 Analysis at the Computational and Algorithmic Levels

Fuzzy logic is considered a good model of human ability to compute with words and make decisions in complex situations when Accuracy and Significance are two features of our statements, which are mutually exclusive. Fuzzy logic has been defined as a rigorous logic of vague reasoning [6]. It is based on the theory of Fuzzy sets [7]. A Fuzzy set is different from a classical set because it breaks the law of excluded-middle. In fact, an item may belong to a Fuzzy set and its complement at the same time, and with the same or different degrees of membership. For example, if we consider the set of the days of the weekend, a classical definition of this set will include only Saturday and Sunday. But, what about Friday? According to the classical definition of this set, the weekend starts at the midnight of Friday and stops at the midnight of Sunday. This description does not encompass completely our perception of the weekend, which usually starts on Friday evening, when we give up working. A Fuzzy description of the set of the days of the weekend is better because it will include also Friday, although with a different degree of membership respect to Saturday and Sunday. The degree of membership of an item to a Fuzzy set can be any number included between 0 and 1. This means that Fuzzy logic is an infinite-valued logic. Fuzzy logic is suitable to describe any complex input-output relation of the type cause and effect. We need just to build a Fuzzy Logic System. For this goal, a three-steps procedure must be followed. The first step requires the *granulation* of all the variables. All the possible values for each variable are partitioned in Fuzzy sets. The number, position and shape of Fuzzy sets depend on the context. The second step requires the *graduation* of all the variables. Each Fuzzy set is labelled by an adjective. The third step consists in formulating the Fuzzy rules that describe the cause and effect relations between the input-output variables. In fact, the Fuzzy rules are syllogistic statements of the type If…, Then… involving the adjectives used to label the Fuzzy sets. When there are multiple inputs, the variables are connected through the operators AND, OR, NOT. At the end of this procedure, we have a Fuzzy Logic System that consists of three main elements. First, the Fuzzifier that transforms numerical values of the input variables in Fuzzy sets. Then, the Fuzzy Inference Engine, which is based on the Fuzzy rules and activates specific output Fuzzy sets. Finally, the Defuzzifier that transforms the activated output Fuzzy sets in a numerical value for the output variable. This means that a Fuzzy Logic System is a predictive tool or a decision support system for the particular phenomenon it describes.

Why is Fuzzy logic a good model of human ability to compute with words? Because there are structural and functional analogies between a Fuzzy Logic System and the human nervous system [8]. In particular, the human sensory system can be described as a collection of Fuzzifiers. We have photoreceptors to detect visible photons, chemoreceptors to probe many chemicals, mechanoreceptors to sense mechanical stimuli, and thermoreceptors to probe thermal stimuli. The multiple information of a stimulus, i.e. its modality (M), intensity (I_M), spatial distribution ($I_M(x,y,z)$) and time evolution ($I_M(x,y,z,t)$), is encoded hierarchically by each element of the sensory system. As an example, let us focus on the visual sensory system. It has a hierarchical structure. At the bottom level, it has four types of photoreceptor proteins; each one has its own absorption spectrum in the visible, although they all have retinal as chromophore. However, the four proteins differ in the amino-acidic composition of the pocket embedding the retinal. At an upper level, we have four types of photoreceptor cells. Each one contains millions of replicas of one particular photoreceptor protein. At the highest level, we have millions of replicas of the four types of photoreceptor cells spread on a tissue that is the retina located inside our eyes. One type of photoreceptor cell, the so-called rod, is distributed on the periphery of the retina and it works when there is light of low intensity. The three other types of cells, the so-called cones, are concentrated in the center of the retina, the fovea, and they allow us to distinguish the colors. How is it possible? The three types of photoreceptor proteins play as three molecular fuzzy sets and the information regarding the modality of the light stimulus is encoded as degree of membership of the light stimulus to the molecular fuzzy sets; in other words, the modality is encoded as Fuzzy information at the molecular level ($\bar{\mu}_{ML}$). The three types of photoreceptor cells play as cellular Fuzzy sets and the information regarding the intensity of the light stimulus is encoded as degree of membership of the light stimulus to the cellular Fuzzy sets; in other words, it is encoded as Fuzzy information at the cellular level ($\bar{\mu}_{CL}$). Finally, we have an array of cellular Fuzzy sets on the retina, and the information regarding the spatial distribution of the light stimulus is encoded as degree of membership to the array of the cellular Fuzzy sets; in other words, it is encoded as Fuzzy information at the tissue level ($\bar{\mu}_{TL}$). At the end, a matrix of data, reproducing the distribution of the photoreceptor cells on the fovea, represents the codification of the overall information of the light stimulus, i.e. the Fuzzy information at the tissue level. Each term of the matrix is the product of two contributions:

$$(\bar{\mu}_{ML} \times \bar{\mu}_{CL}) = (\Phi_{PC} I_{0,\lambda}(1 - 10^{-\varepsilon Cl})) \tag{1}$$

In Eq. (1), Φ_{PC} is the photochemical quantum yield for the photo-isomerization of retinal, ε is its absorption coefficient, C its concentration, l is the optical path within the cell, and $I_{0,\lambda}$ is the intensity of the light stimulus at wavelength λ.

To prove if this computational and algorithmic analysis is effective in describing the way we distinguish colors, it is necessary to implement Eq. (1). For its implementation we have used a collection of direct, thermally reversible, photochromic compounds, as explained in the next paragraph.

4 Implementation of Biologically Inspired Photochromic Fuzzy Logic (BIPFUL) Systems

A direct thermally-reversible photochromic species is a compound that, in absence of solar radiation, it exists in a structure that is usually uncolored, because it absorbs only in the UV. An example is the structure of the spirooxazine labelled SpO in Fig. 1. Upon UV irradiation, the structure changes. MC is produced. MC has an absorption band also in the visible, centered at 600 nm in acetonitrile, conferring blue color to the solution. MC is thermally metastable. Therefore, if the UV irradiation is discontinued, the color bleaches at room temperature.

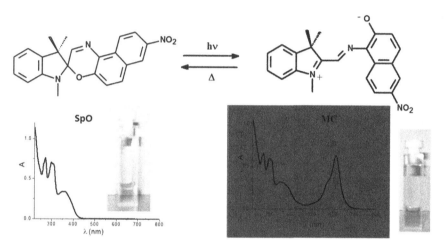

Fig. 1. An example of direct thermally reversible photochromism: SpO is the uncolored form that transforms into MC upon UV irradiation. When SpO converts to MC, the solution turns to blue.

Five photochromic compounds, whose structures are shown in Fig. 2, have been synthesized [9]. Each compound has its own spectral profile in the UV (see panel A in Fig. 2). Moreover, each compound produces a specific color upon UV irradiation and a specific band in the visible region (see panel B). By mixing two or more photochromic compounds, we wanted to obtain systems suitable to distinguish the three UV regions (UVA, UVB and UVC) depending on the color they produce. To prepare such systems some criteria have been followed. First, the absorption bands of the uncolored forms have been conceived as input Fuzzy sets. Second, the absorption bands of the colored forms play like output Fuzzy sets. Third, Eq. (2) represents the degree of membership of UV radiation intensity $I_{0,\lambda}$ to the input Fuzzy set for the i-th photochrome ($\varepsilon_{UV,i}$ is its absorption coefficient at wavelength λ, and $C_{0,i}$ is its analytical concentration in solution):

$$(\mu_{UV,i}) = \Phi_{PC,i} I_{0,\lambda}(1 - 10^{-\varepsilon_{Un,i}C_{0,i}l}) \tag{2}$$

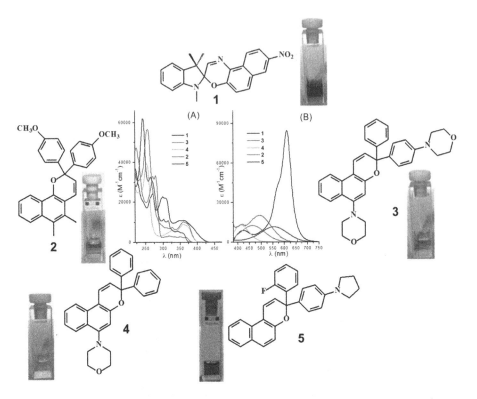

Fig. 2. Structures, spectra and colors of the five photochromic compounds used to implement BIPFUL systems. (Color figure online)

Fourth, Eq. (3) describes the activation of the output Fuzzy set for the i-th photochrome ($\varepsilon_{Co,i}(\lambda_{an})$ is the absorption coefficient of the colored form at the wavelength of analysis λ_{an}, and $(1/k_{\Delta,i})$ is the lifetime of colored species):

$$A_{Co,i} = \frac{\varepsilon_{Co,i}(\lambda_{an})}{k_{\Delta,i}} \mu_{UV,i} \tag{3}$$

Fifth, the equation suitable to predict the color of the solution when we have N photochromic species is:

$$A_{exp} = \sum_{i=1}^{N} A_{Co,i} \tag{4}$$

Equation (4) expresses the sum of the contribution of activated output Fuzzy sets for all the photochromic compounds present into the system.

Based on these five criteria, we have found that many ternary, quaternary and systems containing all the five photochromic compounds of Fig. 2 are effective in distinguishing the UV regions and their intensities [10]. They have been called Biologically Inspired Photochromic Fuzzy Logic (BIPFUL) systems. One of the best BIPFUL systems is a quaternary solution containing compounds **1** (at concentration of 5.2×10^{-5} M), **2** (at 1.4×10^{-4} M), **4** (at 7.4×10^{-5} M), and **5** (at 1.4×10^{-4} M). Such system becomes green upon UVA, grey upon UVB, and orange upon UVC. All these colors can be predicted by using Eqs. (2)–(4). The same equations are useful to predict the color when the BIPFUL system is irradiated by many UV frequencies. The performances of our BIPFUL systems could be extended to solid cellulose supports like white paper [9].

5   The Fuzziness of Molecules

In the previous paragraphs, we have seen that a strategy to implement Fuzzy logic consists in using UV-visible absorption bands as Fuzzy sets and the photochromism phenomenon as Fuzzy Inference Engine. But, there exists another way for implementing Fuzzy logic. It exploits the set of conformers of a compound as Fuzzy set. In fact, the properties of a set of conformers is context-dependent like the properties of a Fuzzy set, and like the meaning of a word in natural language.

When the spirooxazine SpO, shown in Fig. 1, transforms in MC by UV irradiation, it exists as a set of conformers. In fact, MC has a flexible molecular skeleton. The number and type of conformers depend on the context [11]. For example, by heating, the number of conformers increases and their lifetimes shorten. On the other hand, if a zwitterionic amino-acid, like glycine, is added to the solution containing MC, the distribution of

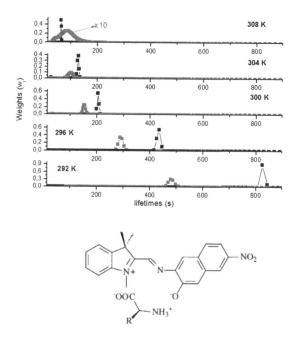

Fig. 3. Distributions of the lifetimes of MC conformers in absence (grey points) and in the presence (dashed points) of glycine at different temperatures. On top, the interplay between glycine and a conformer of MC is depicted.

conformers is significantly shrunk. This is apparent on the top part of Fig. 3 wherein the distributions of conformers in absence of glycine (grey points) and in the presence of glycine (black points) are shown at different temperatures.

The fuzziness of MC is clearly context-dependent. The interplay between glycine and MC (represented on the bottom part of Fig. 3) selects the kind and number of MC conformers.

This second strategy for implementing Fuzzy sets is particularly relevant when we consider macromolecules, especially intrinsically unstructured proteins [12], because they exist under many conformers. It has been demonstrated that regions or whole proteins lacking a well-defined 3D structure can crucially contribute to recognition functions [13]. Structural ambiguity can account for different activities of the same protein, for example, moonlighting [12]. Therefore, structural fuzziness seems to play a relevant role in the molecular events occurring within cells and Fuzzy logic might be a good model for describing how cells and cellular compartments make decisions.

6   Mimicking Micro-electronics

There is also a third strategy for implementing Fuzzy logic by chemicals. It mimics the best way for implementing Fuzzy logic by using micro-electronics [14]. In order to process discrete logics (binary, ternary, and so on), the electronic circuits have to convey and process electrical signals that vary steeply, in sigmoid manner; on the other hand, the best accomplishments of Fuzzy Inference Engines have been achieved so far by analogue electronic circuits that are based upon signals varying smoothly, for instance, in hyperbolic manner.

It has demonstrated [15] that the chromogenic properties of the spirooxazine SpO (see Fig. 1 for its structure) can be exploited to implement Fuzzy Logic Systems whose rules contain all the fundamental Fuzzy logic operators: AND, OR, NOT. To quantify the information a chromogenic species may send to human eyes, it is useful to measure its Colorability (C):

$$C = x_{Col} \log_2 x_{Col} - x_{Unc} \log_2 x_{Unc} + y_{Col} \log_2 y_{Col} - y_{Unc} \log_2 y_{Unc} + z_{Col} \log_2 z_{Col} - z_{Unc} \log_2 z_{Unc} \qquad (5)$$

In Eq. (5), the definition of the Colorability grounds on the Shannon's theory of Information [11]. The terms x, y and z are the chromaticity coordinates values for the uncolored (Unc) and colored (Col) states, respectively. The values of the chromaticity coordinates can be changed in continuous manner by proper selection of the stimulus and its magnitude. We already know that SpO is photochromic. Upon UV irradiation in acetonitrile, we obtain a blue color. If the irradiation is carried out in the presence of one equivalent of protons, the solution becomes orange. The same orange can be achieved in the presence of one equivalent of Al^{+3}. On the other hand, in the presence of one equivalent of Cu^{+2} ions, the solution becomes yellow. More colors can be achieved by adding the ions in analogue manner. The trends of the colorability C as function of the equivalents of H^+ and Cu^{+2} or H^+ and Al^{+3} make them suitable to implement Fuzzy Logic Systems. In particular, the asymmetric shape of the function expressing the dependence of C on H^+ and Cu^{+2} equivalents allows to formulate Fuzzy rules involving just the AND operator. On the other hand, the symmetric shape of the function expressing the dependence of C on H^+ and Al^{+3} equivalents allows to formulate Fuzzy rules involving also the OR and NOT operators. Examples of Fuzzy rules are reported in Table 1.

Table 1. Examples of Fuzzy rules for two Fuzzy Logic Systems having the number of equivalents of H^+, Cu^{2+}, Al^{3+} as inputs and the colorability (C) as output. The symbols VL, L, M, H, VH stand for Very Low, Low, Medium, High, Very High. They are labels of Fuzzy sets.

IF	N° equiv. H^+	Operator	N° equiv. Cu^{2+}	THEN	C
	VL	AND	VL		H
	M	AND	VL		VL
	VL	AND	VH		VH
IF	N° equiv. H^+	Operator	N° equiv. Al^{3+}	THEN	C
	M	AND	NOT VL		L
	H	OR	H		L
	VH	OR	VH		L

7 Conclusions

The development of Chemical Artificial Intelligence can be useful to tackle the Complexity Challenges. In fact, the analysis of the human nervous system and human intelligence at the computational and algorithmic levels promise to unveil secrets of Natural Complexity. For instance, the results described in this work demonstrate that complex systems encode and process information hierarchically.

The analysis of human nervous system at the implementation level and the attempts of mimicking the performances of human intelligence by chemicals and chemical reactions will boost the development of a new generation of computing machines. These new computing machines will ground on wetware and they will look more like a brain rather than like an electronic computer. These new computational machines promise to be useful for facing the problem of recognizing variable patterns, that is one challenge in the field of Computational Complexity.

References

1. Weaver, W.: Science and complexity. Am. Sci. **36**, 536–544 (1948)
2. Rozenberg, G., Bäck, T., Kok, J.N.: Handbook of Natural Computing. Springer, Berlin (2012)
3. Gentili, P.L.: Small steps towards the development of chemical artificial intelligent systems. RSC Adv. **3**, 25523–25549 (2013)
4. Gallistel, C.R., King, A.P.: Memory and the Computational Brain. Wiley, Chichester (2010)
5. Marr, D.: Vision - A Computational Investigation into the Human Representation and Processing of Visual Information. The MIT Press, Cambridge (2010)
6. Zadeh, L.A.: A new direction in AI. AI Mag. **22**, 73–84 (2001)
7. Zadeh, L.A.: Fuzzy sets. Inf. Control **8**(3), 338–353 (1965)

8. Gentili, P.L.: The human sensory system as a collection of specialized fuzzifiers: a conceptual framework to inspire new artificial intelligent systems computing with words. J. Intell. Fuzzy Syst. **27**, 2137–2151 (2014)
9. Gentili, P.L., Rightler, A.L., Heron, B.M., Gabbutt, C.D.: Extending human perception of electromagnetic radiation to the UV region through biologically inspired photochromic fuzzy logic (BIPFUL) systems. Chem. Commun. **52**, 1474–1477 (2016)
10. Gentili, P.L., Rightler, A.L., Heron, B.M., Gabbutt, C.D.: Discriminating between the UV-A, UV-B and UV-C regions by novel Biologically Inspired Photochromic Fuzzy Logic (BIPFUL) systems: a detailed comparative study. Dyes Pigm. **135**, 169–176 (2016)
11. Gentili, P.L.: The fuzziness of a chromogenic spirooxazine. Dyes Pigm. **110**, 235–248 (2014)
12. Tompa, P., Szász, C., Buday, L.: Structural disorder throws new light on moonlighting. Trends Biochem. Sci. **30**, 484–489 (2005)
13. Tompa, P., Fuxreiter, M.: Fuzzy complexes: polymorphism and structural disorder in protein-protein interactions. Trends Biochem. Sci. **33**, 2–8 (2008)
14. Gentili, P.L.: Molecular processors: from qubits to fuzzy logic. ChemPhysChem **12**, 739–745 (2011)
15. Gentili, P.L.: The fundamental Fuzzy logic operators and some complex Boolean logic circuits implemented by the chromogenism of a spirooxazine. Phys. Chem. Chem. Phys. **13**, 20335–20344 (2011)

Mathematical Modeling in Systems Biology

Olli Yli-Harja[✉], Frank Emmert-Streib, and Jari Yli-Hietanen

Tampere University of Technology, Tampere, Finland
olli.yli-harja@tut.fi

Abstract. In this opinion paper we describe how mathematical models can serve as the foundation for communication within multidisciplinary research teams by providing a useful joint context. First we consider the role of mathematical modeling in systems biology in the light of our experiences in cancer research and other biological disciplines in the realm of big data. We examine the methodologies of machine learning, observing the differences between the modeling approach and the black box approach. Next, we consider the role of mathematical models in natural sciences, observing three simultaneous goals: prediction, knowledge accumulation, and communication. Finally, we consider the differences of the pathway model and the attractor model in describing genetic networks, and explore the long-standing criticality hypothesis, discussing its value in multidisciplinary research.

Keywords: Computational biology · Systems biology · Mathematical modeling

1 Introduction

The prediction of the report on New Biology [1] is that serious global challenges will arise in the fields of health, environment, energy, and food production during the 21st century. The report makes a recommendation that biology-based solutions to these societal problems should be sought through gaining deeper understanding of biological systems. Equipped with working knowledge of the organizing principles which describe the relation between structure and function in biological systems we will improve our position in predicting, analyzing, and modulating their behavior. To achieve this the report emphasizes scientific integration - catalyzed by new computational tools. This requires a massive sharing of knowledge and views between fields such as biology, mathematics, engineering, physics, chemistry, science education, and computer science. In many individual research teams such a multidisciplinary collaboration is happening already by necessity. When such collaboration becomes a standard and its scale increases, we expect fast development in computational tools intended to support multidisciplinary collaboration for example in data integration, model development for systems biology, and visualization of large masses of data. For example, cancer research in particular touches many disciplines, such as bioinformatics,

© Springer International Publishing AG 2017
F. Rossi et al. (Eds.): WIVACE 2016, CCIS 708, pp. 161–166, 2017.
DOI: 10.1007/978-3-319-57711-1_14

computational biology, complex systems, modeling of biological systems, and systems biology. In this paper we implicate the importance of collaborative learning and co-creation, and ideas drawn from psychology of collaborative learning as a source of guiding principles in the development of the computational knowledge sharing methods. These methods can be built on the foundation laid by the long history of research in computer-supported collaborative learning [2].

1.1 Big Data in Cancer Research

Big data is seen to hold great promise in solving complex problems. Fashionable black box methods, such a deep learning are expected to provide a shortcut to meaningful interpretations [3]. The main attraction in such black box approaches is that its implementation is easy and it doesn't require any context-specific knowledge. The downside is that it doesn't help in building knowledge of the subject area. Admittedly, in classification tasks the black box approach may provide useful results as an explorative tool, but when we try to observe what has been learned, we see only ad hoc parameters such as number of levels etc.

The sample size requirements for black box methods will always outgrow the number of available samples. Take for example prostate cancer: considering the thousands of free parameters in a black box model, even 3.5 Billion samples, the whole human male population is not enough. With extreme effort and expense it may be possible to collect and analyze say 50 samples for a study. This may be enough to answer a simple, well defined question, but considering the complexity and diversity of the disease, it is not enough to form a clear picture of the situation. Thus, the blind use of black box methods will eventually be a dead end in analyzing big data in cancer research because we cannot build on what has been learned. Instead, every black box effort begins from the same starting point: zero knowledge [4].

In order to build knowledge of a disease, the use of modeling approach is needed. This involves building mathematical models that capture meaningful context information, capable of representing the existing domain knowledge. For example, we have built such a model which interprets TCGA (The Cancer Genome Atlas) data on 8 cancers [5]. Our model utilizes publicly available information in pathway databases, forming an intermediate layer between genomic information and the disease. The pathway layer facilitates a phenotypically meaningful interpretation of the genomic data and allows useful conclusions to be made. As opposed to the black box methods, here we can observe what has been learned from the data, and we can use it as a starting point for further, improved models.

1.2 Characteristics of Mathematical Models and the Scientific Method in Biology

Although the pathway model described above is extremely simple, just an enrichment mapping from genomic data to the pathways combined with a mapping from pathways to the diseases, it has the characteristics of a mathematical model.

It aims at prediction of experimental results and generalization to unseen data. The model is meaningful for cancer researchers, and therefore ways to improve it can be planned. For example the question can be raised, how to incorporate other generic biological information available in databases without complicating the model too much in relation with the available data. Another good question, in the spirit of the report on New Biology [1] is, how we can use the model in bridge-building between scientific disciplines. Note that both of these questions would be meaningless in the context of black box methods.

The scientific method relies on prediction based on mathematical models, and refining those models based on controlled experiments. Let us first consider biological experiments. The scientific method assumes our target, the biological system, to be isolated from its environment in the experiment. However, this can not usually be the case since living systems depend on their environment. For example, a tumor, when removed from a patient is cut off from blood circulation, which fundamentally alters its metabolism. What we can do about this is to try to record relevant details of the experimental environment as well as possible. Similar problems arise with controllability. For example, repeating experiments concerning one metabolic pathway in yeast requires access to the exactly same yeast in exactly same conditions, and still? Since yeast is not a simple device such as an alarm clock, we cant be sure whether some interfering biological processes are present in our experiment.

Another important issue that arises when the scientific method is applied in biology is that of Occam's razor. We choose our predictive model from the high-dimensional space of all possible models. Then, according to the scientific method, we refine this model depending on how its predictions agree with experiments. Since there is an infinity of model refinements to choose from, we have to regularize somehow. Typically regularization is performed by favoring simplicity, although another meaningful choice could be computation time or expense, in the era of computational science [6].

We can safely agree that biological systems are complex, still we should follow Einstein's advise "In the limited nature of the mathematically existent simple fields and simple equations possible between them, lies the theorist's hope of grasping the real in all its depth" [7]. We cannot expect the simplest solution to be more true than any other solution that produces the same predictions, but we can expect it to be more meaningful for us. Although, in aesthetic sense simplicity and beauty may go hand in hand, we still can't equate beauty with truth. We propose that the reason for favoring the simplest solution lies in the limitations of our capacity, in understanding and mainly in communication. The value of simplest equations arises from their efficiency in serving as a joint context for communication. This applies to one to one discussions, as well scientific publishing. Here, the aesthetic value of the model starts to play a role as a factor increasing its popularity.

Let us imagine a discussion between a biologist and an engineer:

Engineer: "Hi, let's work together on mitochondrial diseases!"
Biologist: "Ok, why?"

Engineer: "I can analyze data and I can model things."
Biologist: "Ok, what kind of models?"
Engineer: "Computational models."
Biologist: "Aha, that kind of mice:)"
Engineer: "Hmm - attractor models ..."
Biologist: "Hmm?"
Engineer: "Hm?"
Biologist: "Read these thousand papers and come back next year."

It seems that it could be very useful to have a joint context in the conversation. In order to be practical, such a context shouldn't be extremely complex (like thousand research papers). If the parties could agree about a few concepts, the discussion could have been much more fruitful: for example the concept of a model - what does it represent and what it can do - can it eat cheese, can it be used as a pointer, or does it have predictive value; do we identify attractor with a cell type, or attractive fashion models. Altogether we conclude that justification for the widespread use of Occam's razor arises naturally from the need of joint context in scientific communication, while the reality that we are trying to describe remains overwhelmingly complex, as the case of biology clearly points out.

1.3   Criticality and the Attractor Model

Building a picture of a biological system is extremely tedious, and it would benefit from an efficient guiding paradigm. The key issue is, what are the entities that we talk about in describing research and experiments. Pathways are entities that clearly have a useful function in understanding biological processes. They are much closer to the phenotype than genes are, and meaningful interpretation of e.g. diseases can be achieved by considering pathways [5]. However, pathways connect to each other and although their description casts some light on the structure of biological systems, the functional description based on the pathway concept doesn't seem to be as useful.

The question can be raised, could there be other entities that we could use as models in order to understand, not only the structure, but also the function of biological systems? Which models would tolerate, or even make use of the high connectivity of such systems. We propose to consider the long standing criticality hypothesis by Stuart Kauffman [8] and its interpretation that cell types can be meaningfully modeled as attractors. The appealing features of the attractor model include maximal information transfer [9], maximal flexibility - by maximizing the adjacent possible [10], small world networks [11], power-law distributions [12], etc. Attractors are emergent manifestations of the dynamics of their underlying system structure. Their descriptive power focuses on the phenotypic behavior in biology, e.g. cell types.

Lot of effort has been invested in examining, whether individual biological systems are critical or not, e.g. [13,14]. Instead of revisiting this or trying to evaluate whether the criticality hypothesis is true or not, we propose making use of it and seeing whether or not that leads to predictive value. Some practical attempts to this direction have been made. For example Huang et al. [15] describe their work as follows: "We used gene expression profiling to show that trajectories of neutrophil differentiation converge to a common state from different directions of a 2773-dimensional gene expression state space, providing the first experimental evidence for a high-dimensional stable attractor that represents a distinct cellular phenotype." The attractor model facilitates describing this research setting in one sentence, while trying to describe it within the pathway paradigm would be fruitless.

Another important effort by Heinniemi et al. identifies self-stabilizing expression states with attractors, and making use of human expression data from 166 cell types develops a method for identifying cell fate, and potentially its conversion [16]. Again, without the idea of cell types described as dynamic attractors, it would have been close to impossible to describe the research setting.

We predict that in the future new biological databases will be built based on the organizing principles derived from the attractor model. New computational tools will also be developed that depend on this research paradigm and support it in visualization, viewpoint sharing, and data integration.

1.4   Conclusion

We emphasize the role of mathematical model as a socially constructed entity [17]. Mathematical modeling in systems biology relies on the quality of human and technologically mediated communication, shared problem representation and knowledge building and seamless collaboration in the research teams. We should aim at a scientific breakthrough in the study of collaborative modeling in multidisciplinary team environments, especially in life sciences.

Several simultaneous trends in the scientific world force new competence requirements upon the collaborating scientists: intensifying virtual collaboration, larger multidisciplinary research teams, increasing computation and storage capacity, big data, open science [18], improving scientific visualization, and even game applications in research. Modeling complex problems is itself becoming a scientific problem of increasing value. We expect consortia to be established with experts on collaborative learning, co-creation, cognitive science, computational and modeling tools, and scientists working on various aspects of systems biology. The basic requirement within such consortia is efficient communication, and we propose that mathematical models can serve as the foundation of shared context.

References

1. Committee on a New Biology for the 21st Century. Report on new biology. The National Academies Press, Washington (2009)
2. Stahl, K., Suthers, C.-S.: collaborative learning: an historical perspective. In: Sawyer, R.K. (ed.) Cambridge Handbook of the Learning Sciences, pp. 409–426. Cambridge University Press, Cambridge (2006)
3. LeCun, Y., Bengio, Y., Hinton, G.: Deep learning. Nature **521**, 436–444 (2015). 15

4. Yli-Hietanen, J., Ylip, A., Yli-Harja, O.: Cancer research in the era of next-generation sequencing and big data calls for intelligent modeling. Chin. J. Cancer **34**, 12 (2015)

5. Ylip, A., Yli-Harja, O., Zhang, W., Nykter, M.: Characterization of aberrant pathways across human cancers. BMC Syst. Biol. **7**, S1 (2013)

6. Yli-Harja, O., Ylip, A., Nykter, M., Zhang, W.: Cancer systems biology: signal processing for cancer research. Chin. J. Cancer **30**, 221–225 (2011)

7. Einstein, A.: Essays on Science, p. 19. Philosophical library, New York (1934)

8. Kauffman, S.: Metabolic stability, epigenesis in randomly constructed genetic nets. J. Theor. Biol. **22**(3), 437–467 (1969)

9. Chowdhury, S., Lloyd-Price, J., Smolander, O.-P., Baici, W., Hughes, T., Yli-Harja, O., Chua, G., Ribeiro, A.: Information propagation within the Genetic Network of Saccharomyces cerevisiae. BMC Syst. Biol. **4**, 143 (2010)

10. Kauffman, S.: Investigations. Oxford University Press, Oxford (2000)

11. Watts, D., Strogatz, S.: Collective dynamics of 'small-world' networks. Nature **393**, 440–442 (1998)

12. Bak, P., Tang, C., Wiesenfeld, K.: Self-organized criticality: an explanation of $1/f$ noise. Phys. Rev. Lett. **4**, 59 (1987)

13. Nykter, M., Price, N., Aldana, M., Ramsey, S., Kauffman, S., Hood, L., Yli-Harja, O., Shmulevich, I.: Gene expression dynamics in the macrophage exhibit criticality. PNAS **105**(6), 1897–1900 (2008)

14. Rm, P., Kesseli, J., Yli-Harja, O.: Perturbation avalanches and criticality in gene regulatory networks. J. Theor. Biol. **242**, 1 (2006)

15. Huang, S., Eichler, G., Bar-Yam, Y., Ingber, D.: Cell fates as high-dimensional attractor states of a complex gene regulatory network. Phys. Rev. Lett. **94**, 128701 (2005)

16. Heinniemi, M., Nykter, M., Kramer, R., Wienecke-Baldacchino, A., Sinkkonen, L., Zhou, J., Kreisberg, R., Kauffman, S., Huang, S., Shmulevich, I.: Gene-pair expression signatures reveal lineage control. Nature Methods **10**, 577–583 (2013)

17. Giere, R.: Explaining Science. A Cognitive Approach. Chicago University Press, Chicago (2010)

18. Emmert-Streib, F., Dehmer, M., Yli-Harja, O.: Against dataism and for data sharing of big biomedical and clinical data with research parasites. Front. Genet. **7**, 154 (2016)

Synchronization in Near-Membrane Reaction Models of Protocells

Giordano Calvanese[1], Marco Villani[1,2(✉)], and Roberto Serra[1,2]

[1] Department of Physics, Informatics and Mathematics,
University of Modena and Reggio Emilia, Modena, Italy
`calvanese.giordano@gmail.com`,
`{marco.villani,rserra}@unimore.it`
[2] European Centre for Living Technology, Ca' Foscari University, Venice, Italy

Abstract. In this paper a new model of growing and dividing protocells is described, whose main features are (i) an autocatalytic set of "genetic memory molecules" (GMMs) whose reactions happen in a thin aqueous phase shell near the membrane and (ii) a lipid container that grows according to the amphiphilic production stimulated by the GMMs. Synchronization occur when the container growth rate is equal to the GMMs self-replicative one: the behavior of this model is compared with a previous version where reactions occur in the whole internal aqueous volume. Analytical results and simulations has shown that synchronization emerges in both models for the same set of kinetic equations, the main difference being only in the time scale of the process. Moreover the introduction of finite rates in the transmembrane diffusion permits the emergence of synchronization for a significantly wide set of parameters, enough to allow the protocell evolvability (defined as the capability of cumulating novelties, by maintaining the already present capabilities).

1 Introduction

The term "protocell" is used in a loose sense to denote a class of (still hypothetical) entities that are similar to, but much simpler than present-day cells (Rasmussen et al. 2004, 2008). They should be able to grow and to divide, giving birth to offsprings that are similar but not necessarily identical to their parents. Different individuals should reproduce at possibly different rates, thereby allowing selection to occur. The protocells discussed here should not be confused with the "minimal cells" that have been synthesized by simplifying the genome of existing microorganisms (Gibson et al. 2010), since they should be built starting from non-living components. While the relevance of this research for the problem of the origin of life is apparent, in this paper we will be concerned with the behavior of possibly synthetic protocells, not referring to plausible scenarios for abiogenesis.

While a number of alternative protocell architectures have been proposed, most of them are based on a lipid container (e.g. a vesicle), with an internal aqueous phase and a membrane formed by a double layer of amphiphilic molecules. Moreover, it is assumed that there is a set of collectively self-replicating molecules, that can be referred to as "replicators" or "genetic molecules". Let us call for brevity "key reactions" those

© Springer International Publishing AG 2017
F. Rossi et al. (Eds.): WIVACE 2016, CCIS 708, pp. 167–178, 2017.
DOI: 10.1007/978-3-319-57711-1_15

that are involved either in the growth of the container or in the duplication of the genetic material (Villani et al. 2016). In general, one can distinguish models where the key reactions take place inside the membrane (*Surface Reaction Models* or SRMs) (Serra et al. 2007) from those models where they take place in the homogeneous internal aqueous phase (*Internal Reaction Models*, shortly IRMs) (Carletti et al. 2008; Filisetti et al. 2010). Simplifying assumptions about the concentration profiles inside the aqueous phase are often considered, the simplest one being that of homogeneous concentrations.

However, there is a problem with IRMs that is often overlooked: suppose indeed that the vesicle is large enough so that composition fluctuations are small in a volume of the same size as that of the internal phase. If the protocells are generated by some spontaneous process taking place in a homogeneous environment, then the internal composition of all the protocells will be very similar to each other, and to the external environment. So essentially the same reactions take place in each protocell, and in the environment – and there is really no need to have a closed compartment. Since all life forms are based upon cells, these must instead be very important, probably from the very beginnings of life. A possible solution to this problem is that of assuming that the initial protocells were so small that the fluctuations were large in their very small volumes, so their compositions are different and selection can take place (Serra et al. 2014).

However, there is an alternative possibility, which would hold also in the case where the initial vesicles were quite large: it is possible that the key reactions take place inside the vesicle, but only very close to its membrane, which is supposed to provide direct catalytic activity or to give rise to a local environment that favours those reactions. We will refer to these architectures as *Near-Membrane Reaction Models*, shortly NMRMs (see Fig. 1).

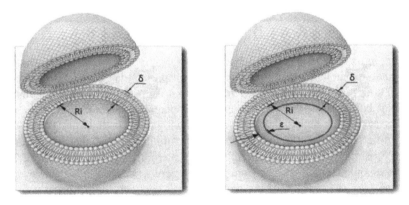

Fig. 1. Schematic representation of an Inner Reaction Model (IRM - left) and of a Near Surface Reaction Model (NMRM - right)

While they resemble IRMs, where the internal phase is supposed to be homogeneous, the main difference is that in this case the production of new self-replicators takes place in a narrow spherical shell close to the membrane. One might argue that the same catalytic activity takes place on the outer side of the membrane but, if the external volume is much larger than the internal one, the reactions products will be quickly diluted on the outside, while this will not happen inside the protocell, if the membrane is impermeable. Therefore the internal chemical composition may be different from the external one.

In this paper we will introduce an abstract Near-Membrane Reaction Model (NMRM in the following) and we will then address the important problem of synchronization.

In a series of papers, we (Serra et al. 2007; Carletti et al. 2008; Filisetti et al. 2008; 2010; Villani et al. 2014) and others (Munteanu et al. 2007) have drawn attention on the importance of synchronization between the replication rates of the "container" and of its genetic material, that is obviously a necessary condition for sustained growth of a protocell population. Using fairly abstract models, it had been possible to prove that, under a wide set of hypotheses, such synchronization spontaneously emerges, generation after generation both in Internal Reaction Models and in Surface Reaction Models.

In the case of NMRMs, the replication of the genetic molecules takes place only in a fraction of the internal volume, so the self-replicators can then undergo dilution during the growth of the protocell - and this might affect synchronization. However, we show here that synchronization is achieved also in this case under a broad range of assumptions concerning the type of equations and the sets of parameter values that describe (i) the interactions among the replicators and (ii) the interactions of some replicators with the lipid container.

We will then compare the behaviour of IRMs and NMRMs using simplified deterministic dynamical models, and assuming that the concentrations of the self-replicators are the same in every point of the internal aqueous phase for IRMs and NMRMs, while only a part of the self-replicators in the NMRM case participate to the reaction processes (that is, the part in the spherical shell close to the membrane). Diffusion in the internal phase is supposed to be instantaneous, while transmembrane diffusion of the precursors of the genetic molecules and of the amphiphiles can be either instantaneous or ruled by a finite diffusion coefficient. In the first case, some results have been obtained using analytical methods while in the second case all the results are based upon simulations.

2   The Models

2.1   Inner- and Near Membrane-Reaction Models

The systems described in this paper require the modelling of two main interacting subsystems: (i) the chemical reactions dynamics happening inside the container and (ii) the container itself.

The world that the protocells inhabit is defined by chemistry, that is, a set of chemical species and catalysed reactions. In the following we use several "random generated" chemistries, which can differ in characteristics as number of species and number of reactions. Moreover, the chemistries are generated at random, so the same molecule can catalyse a certain reaction in one chemistry and not in another one. Of course, in the real world there is just one "chemistry", but we consider ensembles of randomly generated cases in order to look for generic properties.

Let $\mathbf{Y} = \{X_1, X_2,...,X_N\}$ the whole set of N chemical species composing the chemistry (X_i being the amount – the number of moles - of the species X_i, and $[X_i]$ its concentration). Some of these chemicals catalyse the formation of (other) chemicals and/or the production of the protocell container, by using the following rules:

$$\begin{cases} \frac{d[C]}{dt} = \alpha_1 [P_C]^k [X_1]^\gamma + \ldots + \alpha_n [P_C]^k [X_n]^\gamma \\ \frac{d[X_i]}{dt} = \eta_{i,1} [P_{X_i}]^k [X_1]^v + \ldots + \eta_{i,n} [P_{X_i}]^k [X_n]^v \quad i = 1,\ldots,n \end{cases} \tag{1}$$

where P_c and P_{Xi} are respectively the substrates (the "precursors") of the amphiphilic molecules forming the protocell vessel and the precursors of the chemicals that, placed inside the vessel, compose the "metabolism" of the protocell itself. The fact that (a subset of) these chemicals are able to catalyze the formation of (other) chemicals and/or influence the growth of the protocell container allows them to heavily affect the behavior of the whole system: therefore we can refer to these chemical species as a sort of "genetic memory molecules" (GMM in the following) of the system. The matrix η_{ij} and the vector α_i define respectively the interaction between the GMMs and their effect on the protocell growth; the exponents k, γ and v can differ from 1 in order to take into account non-linear processes.[1]

We can change variables, from concentrations to quantities, by explicitly introducing the volume of the protocell V_{int} and the volume V_{eff} where the main reactions happen:

$$\begin{cases} \frac{dC}{dt} = \alpha_1 P_C^k X_1^\gamma \frac{V_{eff}}{V_{int}^{\gamma+k}} + \ldots + \alpha_n P_C^k X_n^\gamma \frac{V_{eff}}{V_{int}^{\gamma+k}} \\ \frac{dX_i}{dt} = \eta_{i,1} P_{X_i}^k X_1^v \frac{V_{eff}}{V_{int}^{v+k}} + \ldots + \eta_{i,n} P_{X_i}^k X_n^v \frac{V_{eff}}{V_{int}^{v+k}} \quad i = 1,\ldots,n \end{cases} \tag{2}$$

The protocell volume is determined by the shape of the membrane and by the amount of amphiphilic molecules: if during growth the protocell maintains more or less the same shape protocell volume and amount of amphiphiles are deterministically linked, and we can compute the protocell volume by knowing the amphiphiles quantity. For example, in case of spherical (turgid) protocell surrounded by a membrane having a thickness of δ and a density of amphiphiles equal to ρ we have (Villani et al. 2014):

---

[1] In order to simplify the discussion we use the same exponent k to describe the effect of the concentrations of both kind of precursors on the order of their respective equations - this simplification nevertheless does not affect the main conclusions of the paper. Another simplification regards the exponents v (equal for all GMMs) and γ (equal for all interactions among GMMs and the precursors of the container): indeed, we are taking into consideration chemical reactions having not too different properties.

$$V_{eff}(C) = V_{int} = \frac{1}{6\sqrt{\pi}} \left(\frac{C}{\rho\delta}\right)^{\frac{3}{2}} \tag{3}$$

whereas if reactions happen only within a distance ε from the inner surface of the membrane we have:

$$V_{eff}(C) = \frac{C}{\rho\delta}\varepsilon \tag{4}$$

We assume that both kinds of precursors can cross the membrane (that is, that they are small molecules or have some affinities with the membrane). Usually the external environment is enormously wider than the protocell size, so the processes here happening cannot change it in substantial way: in other words, the external concentrations do not appreciably change in time (they are constant). If we assume fast transmembrane diffusion, the same holds also for the internal concentrations, and we can simplify Eq. 2 as follows[2]:

$$\begin{cases} \frac{dC}{dt} = \alpha_1 X_1^\gamma \frac{V_{eff}}{V_{int}^\gamma} + \ldots + \alpha_n X_n^\gamma \frac{V_{eff}}{V_{int}^\gamma} \\ \frac{dX_i}{dt} = \eta_{i,1} X_1^\gamma \frac{V_{eff}}{V_{int}^\gamma} + \ldots + \eta_{i,n} X_n^\gamma \frac{V_{eff}}{V_{int}^\gamma} \quad i = 1,\ldots,n \end{cases} \tag{5}$$

Finally, we assume that when the amphiphiles quantity reaches the threshold value θ the membrane becomes unstable and the protocell splits in two parts[3], from which the process can restart (Villani et al. 2014, 2016).

Note that (without constraints) at each generation the protocell population doubles its elements, a process that leads to an exponential growth; at the same time the inner dynamics – which influences the duplication time - is ruled by the kinetic coefficients in Eq. 1, and it could assume linear, sublinear or supralinear behaviors.

2.2   Releasing the Hypothesis of Fast Transmembrane Diffusion

The exchange of materials between the protocell and its environment is a very delicate process: indeed, in our vision it plays a crucial role in assuring the protocells' evolvability. Unfortunately, we will see (in Sect. 3) that the fast cross membrane diffusion hypothesis may have important consequences. In order to consider a more realistic model with finite membrane crossing rates we can take into account the passive transport of materials, which is driven by the gradient between the internal and external chemical concentrations (Fick's law):

---

[2] In Eq. 5 we set $\alpha'_i = \alpha(P_C/V_{int})^k$, and rename α'_i again in α_i in order to avoid an excessive proliferation of symbols. The same holds for $\eta_{i,j}$.

[3] The offspring do not have necessarily the same size: indeed, the preservation of the protocells' size among the different generations is assured by the fixed amount of the amphiphiles at the division event.

$$\frac{dP}{dt} \approx \frac{D_p S([P]_{ext} - [P]_{int})}{\delta} \tag{6}$$

Where dP/dt is the variation of the internal concentration of the chemical, D_p is proportional to its diffusion coefficient through the membrane, S is the membrane surface and δ is its thickness; $[P]_{ext}$ and $[P]_{int}$ are respectively the external and internal concentrations.

This rule, combined with the fact that chemicals can be consumed by the reactions where they are substrates, lead to a slightly different model:

$$\begin{cases} \frac{dC}{dt} = \alpha_1 P_C^k X_1^\gamma \frac{V_{eff}(C)}{V_{int}^{\gamma+k}(C)} + \ldots + \alpha_n P_C^k X_n^\gamma \frac{V_{eff}(C)}{V_{int}^{\gamma+k}(C)} \\ \frac{dX_i}{dt} = \eta_{i,1} P_{X_i}^k X_1^v \frac{V_{eff}(C)}{V_{int}^{v+k}(C)} + \ldots + \eta_{i,n} P_{X_i}^k X_n^v \frac{V_{eff}(C)}{V_{int}^{v+k}(C)} \quad i=1,\ldots,n \\ \frac{dP_C}{dt} = \frac{D_{P_C} S(C)}{\delta} \left([P_C]_{ext} - \frac{P_C}{V_{int}(C)} \right) - \frac{dC}{dt} \\ \frac{dP_{C_i}}{dt} = \frac{D_{P_{X_i}} S(C)}{\delta} \left([P_{X_i}]_{ext} - \frac{P_{X_i}}{V_{int}(C)} \right) - \frac{dX_i}{dt} \qquad i=1,\ldots,n \end{cases} \tag{7}$$

that maintains the more interesting behaviors of the previous ones, by adding – as we will see – some interesting properties.

3 Results

3.1 Fast Cross Membrane Diffusion: Equivalence of Inner- and Near Membrane-Reaction Models

In the case of only one GMM, and with the hypothesis of fast cross membrane diffusion, we can analytically prove that the IRM and NMRM models are equivalent. Indeed, Eq. 5 becomes:

$$\begin{cases} \frac{dC}{dt} = \alpha X^\gamma \frac{g_{i/n}(C)}{g_i^\gamma(C)} \\ \frac{dX}{dt} = \eta X^v \frac{g_{i/n}(C)}{g_i^v(C)} \end{cases} \tag{8}$$

where the functions

$$g_i(C) = \frac{1}{6\sqrt{\pi}} \left(\frac{C}{\rho\delta} \right)^{\frac{3}{2}} = aC^{\frac{3}{2}} \tag{9}$$

$$g_n(C) = \frac{C}{\rho\delta} \varepsilon = bC \tag{10}$$

describe the volume where reactions happen. So, by using the new coordinate:

$$\frac{d\tau}{dt} := \frac{dC}{dt} = \alpha X^{\gamma} \frac{g_{i/n}(C)}{g_i^{\gamma}(C)} \tag{11}$$

we obtain[4]:

$$\begin{cases} \frac{dC}{d\tau} = 1 \\ \frac{dX}{d\tau} = \frac{\eta}{\alpha} X^{\nu-\gamma} \frac{1}{g_i^{\gamma-\gamma}(C(\tau))} \end{cases} \tag{12}$$

that is an easily solvable system.

By using the mapping techniques shown in (Serra et al. 2007; Carletti et al. 2008) it is possible to prove that a sufficient condition for the system to synchronize is $v < \gamma+1$, that is, that the order of the autocatalytic action of the GMM has to be lower than the order of its action on the container growth (Calvanese 2016).

Moreover, the function $g_{i/n}$ is not present in Eq. 12, indicating that IRM and NSRM models synchronize in the same conditions, all significant changes being simply described by using a different time scale.

3.2 Fast Cross Membrane Diffusion: Competition Among GMMs

So, a single GMM coupled with the protocell membrane can allow the sustainable synchronization of a population of protocells. The same holds also for a set of collectively autocatalytic molecules, able to recruit the materials needed to its growth (a so called Reflexively Autocatalytic Food generated set, or RAF set (Steel 2000; Hordijk and Steel 2004; Villani et al. 2014, 2016), if coupled with the protocell membrane (we can call this synchronising RAF "sRAF" (Villani et al. 2016).

But what happens if independent sRAF sets[5] are present within the same protocell? The simulations shows that, as one might expect, independent sRAFs having the same growth rate can coexist in the same protocell, even if they have different coupling coefficients with the membrane, or even if some of them are not coupled at all (these last sRAFs being a sort of guests of the sRAFs that contribute to the container growth).

However, if the sRAFs do not have exactly the same growth rate, the final fate depends on the order of autocatalysis v. Indeed, as observed also in similar dynamical situations (Smith and Szathmáry 1995), there are three main cases:

1. $v < 1$ the protocell inner dynamics is sublinear: in this case almost all different RAFs can synchronize ("survival of everyone")
2. $v = 1$ the protocell inner dynamics is linear: in this case only the fastest sRAF synchronize: all other sRAFs during the process of duplication dilute, irrespectively with respect to the strength of their coupling with the container. The only exception is the possible presence of an even fastest sRAF that does not interact with the container: in this case during duplication its constituents invade the protocells' inner

---

[4] The symbol $g_{i/n}$ stand for g_i or g_n, depending on the involved model.

[5] Note that a single autocatalytic GMM is a RAF set composed by only one reaction.

space, till the block of the final protocells - an event out of the scope of the model ("survival of the fastest")

3. $v > 1$ the protocell inner dynamics is super linear: in this case only the sRAF initially present with the highest concentrations synchronize: all other sRAFs during the process of duplication dilute, irrespectively with respect to the strength of their coupling with the container ("survival of the first")

These scenarios are not good news for the evolvability of the system. Indeed in many cases (typically all having $v \geq 1$) there is only one final fate: this absence of different asymptotic behaviours totally blocks the progress of evolution, which is based on the success of (some) variations. Moreover, the always possible introduction of novelties (for example because of the occurrence of spontaneous reactions or because the interference of the environment) could introduce (at low concentration) parts of sRAFs having alternatively higher or lower growth rates. In the first case these sRAFs replace the already existing ones (further fixing the situation); in the second case they cannot survive and sooner or later dilute (confirming in such a way the robustness of the already existing situation). No scenarios allows the progressive accumulation of useful behaviours – with the only exceptions (i) of the formation of new branches of the already existing sRAF and (ii) of sRAFs having exactly the same growth rate, a very rare occurrence.[6]

3.3   The Effect of Finite Membrane Diffusion

Indeed, if we analyses the flow of chemicals that can cross the membrane in and out the protocell, we can note that in some cases the rates are unrealistically high; typically, the highest rates are related to the chemicals that are substrates or products of the leading sRAFs.

If we consider a more realistic model that takes into account the passive transport of materials – as described in Sect. 2 – the simulations outcomes change: there are significantly wide ranges of parameters where also reaction orders with $v \geq 1$ can easily support the "survival of everyone" scenario. In particular:

- different sRAFs having dissimilar growth rates can coexist (obviously, provided that the differences among the growth rates are not too high)
- a new sRAF can stably enter into the series of duplicating protocells, even if its initial quantity is (not excessively[7]) small, and even if its growth rate is (not excessively (see Footnote 6)) lower than the growth rates of the already existing RAFs

---

[6] There is an alternative way to escape from this dilemma, that is, the scenario where the new behavior (supported by the chemical activity of the new sRAF) is providing a new functionality to the protocell, in a sense "orthogonal" to the functionality supported by the already present sRAF (that is, the system's reproduction). For example, the resistance to aggressive chemical substances, of to the crowding of the environment: both these alternatives are however out of the scope of the model (see (Villani et al. 2014) for a detailed discussion of this theme).

[7] That is, there are threshold values below which the phenomenon does not happen.

- without replacing the already existing sRAFs, if its growth rate is (not excessively (see Footnote 6)) higher than those of the already existing ones
- the IRM variant allows the coexistence of different sRAFs for differences in growth rate higher than the NSRM[8]

So, the finite membrane diffusion plays a key role in allowing evolvability (defined as the capability of cumulating novelties, by maintaining the already present capabilities) (Figs. 2 and 3).

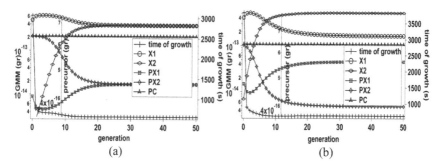

(a)   (b)

Fig. 2. IRMs with $v = 1$ and $k = 1$. (a) The (independent – that is, $\eta_{1,2} = \eta_{2,1} = 0$) GMMs start from different initial quantities ($X_1 = 5e^{-16}$g. and $X_2 = 5e^{-19}$g), and have similar growth rates (cm^3 *s^{-1} g^{-1}): $\eta_{1,1} = \eta_{2,2} = 0,05$. (b) The (independent – $\eta_{1,2} = \eta_{2,1} = 0$) GMMs start from similar initial quantities ($X_1 = X_2 = 5e^{-16}$g.) and have different growth rates (cm^{1,5} * s^{-1}): $\eta_{1,1} = 0,0375$ and $\eta_{2,2} = 0,05$. The finite trans-membrane diffusion of precursors allows their coexistence

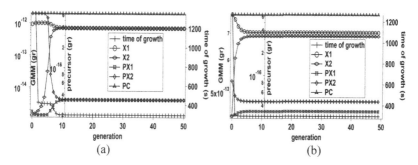

(a)   (b)

Fig. 3. IRMs with $v = 1.5$ and $k = 1$. (a) The (independent – that is, $\eta_{1,2} = \eta_{2,1} = 0$) GMMs start from different initial quantities ($X_1 = 5e^{-16}$g. and $X_2 = 5e^{-19}$g) and have similar growth rates (cm$^{4.5}$ * s$^{-1.5}$ g$^{-1.5}$): $\eta_{1,1} = \eta_{2,2} = 0,158114$. (b) The (independent – $\eta_{1,2} = \eta_{2,1} = 0$) GMMs start from similar initial quantities ($X_1 = X_2 = 5e^{-16}$g.) and have different growth rates (cm$^{4.5}$ * s$^{-1.5}$ g$^{-1.5}$): $\eta_{1,1} = 0,158114$ and $\eta_{2,2} = 0,118114$. The finite trans-membrane diffusion of precursors allows their coexistence

---

[8] The ranges of parameters supporting synchronization is wider for system with $v = 1$ than for systems with $v = 3/2$ (the higher the nonlinearity, the higher the synchronization difficulty). Despite this tendency, both situations own a significant wide range of parameters.

4 Conclusions

It turns out that the behaviour of the Near-Membrane Reaction Models are qualitatively similar to those of the IRMs, although there are some values of the kinetic parameters that lead to extinction (i.e. extreme dilution) in the former case but not in the second. In general, with instantaneous diffusion, one observes coexistence of molecular species that have different replication rates when the kinetics of the self-replicators is sublinear, while in the linear case the fastest one (i.e. the "fittest") prevails. Note that while the replicator equations may be linear, the whole model is definitely nonlinear. If the replicators kinetics are superlinear, one observes that the molecular types with a higher initial concentration have an edge with respect to the other species (the "survival of the first"). The above remarks refer essentially to molecular species that self-replicate individually. When replication involves cooperation between two or more species, they together determine the overall rate of replication - they can survive and synchronize. The presence of several different sets of self-replicating polymers, with different rates, is a desirable feature of protocells, as it might allow them to become more complex and to perform a wider set of functions. It is shown in the paper that the coexistence of different sets of this type, even when the kinetics are linear or superlinear, can be achieved in both IRM and NMRM architectures (the IRM organization being the more flexible one) when the diffusion rate across the membrane is finite – provided that the replication rates of the genetic molecules are not too different from each other. Moreover, a limited analysis is performed concerning the evolvability of the protocell, by assuming that a new molecular species enters the scene when the system has already reached a steady state, and by defining "evolvable" a case when the newcomers can survive together with the pre-existing molecules. It is shown that the introduction of a finite diffusion rate allows the protocell population to evolve, in the sense defined above, also when the replication kinetics is linear or superlinear.

Acknowledgments. The authors gratefully acknowledge the support of the European Centre for Living Technology in Venice and helpful discussions with Stuart Kauffman, Irene Poli, Alessandro Filisetti, Alex Graudenzi, Chiara Damiani, Ruedi Fuechslin and Timoteo Carletti

Appendix A: Symbols and Main Acronyms

Main acronyms	Meaning
SRM	Surface Reaction Model
IRM	Internal Reaction Model
NMRM	Near-Membrane Reaction Model
GMM	Genetic Memory Molecules
RAF	Reflexively Autocatalytic Food generated set
sRAF	A RAF set that allows synchronization if coupled with membrane growth

Symbols	Meaning	First use
X_i $[X_i]$	Respectively, amount (moles) and concentration of species X_i	Equation 1
C $[C]$	Respectively, amount (moles) and concentration of amphiphilic molecules forming the protocell vessel	Equation 1
P_c $[P_c]$	Respectively, amount (moles) and concentration of the precursors of amphiphiles forming the protocell membrane	Equation 1
P_{Xi} $[P_{Xi}]$	Respectively, amount (moles) and concentration of the precursors of the protocell internal chemicals	Equation 1
$[P_C]_{ext}$ $[P_{Xi}]_{ext}$	Respectively, external concentrations of the precursors of amphiphiles and of internal chemicals	Equation 7
V_{int} V_{eff}	Respectively, protocell inner volume and "effective volume" where reactions take place	Equation 2
S	membrane surface	Equation 6
$g_i(C)$ $g_n(C)$ $g_{i/n}(C)$	Volumes where the reactions take place (respectively for the IRM and NSRM models – "i/n" stands for both situations)	Equation 8
η_{ij}	The matrix defining the interaction between the GMMs	Equation 1
α_i	The vector defining the coupling among each internal chemical and the growth of the protocell membrane	Equation 1
κ	The equations order of membrane and internal material growth with respect to the concentrations of the precursors	Equation 1
γ	The equations order of the membrane growth with respect to the internal chemicals concentrations	Equation 1
ν	The exponent giving the equations order of the internal chemicals growth with respect to the internal chemicals	Equation 1
δ	Membrane thickness	Equation 3
ρ	Density of amphiphiles within the membrane	Equation 3
ε	In NSRM model, the thickness of the layer close to the membrane where the reaction take place	Equation 4
D_p	Transmembrane diffusion coefficient for the precursors of the internal chemicals	Equation 6

References

Calvanese, G.: Sincronizzazione in modelli di crescita di protocellule. Degree thesis in Physics, University of Modena and Reggio Emilia (2016)

Carletti, T., Serra, R., Poli, I., Villani, M., Filisetti, A.: Sufficient conditions for emergent synchronization in protocell models. J. Theor. Biol. **254**, 741–751 (2008)

Hordijk, W., Steel, M.: Detecting autocatalytic, self-sustaining sets in chemical reaction systems. J. Theor. Biol. **227**, 451–461 (2004)

Filisetti, A., Serra, R., Carletti, T., Poli, I., Villani, M.: Synchronization phenomena in protocell models. BRL. Biophys. Rev. Lett. **3**(1/2), 325–342 (2008)

Filisetti, A., Serra, R., Carletti, T., Villani, M., Poli, I.: Non-linear protocell models: synchronization and chaos. Europhys. J. B **77**, 249–256 (2010)

Gibson, D.G., Glass, J.I., Lartigue, C., Noskov, V.N., Chuang, R.Y., Algire, M.A., Benders, G.A., Montague, M.G., Ma, L., Moodie, M.M., Merryman, C., Vashee, S., Krishnakumar, R., Assad-Garcia, N., Andrews-Pfannkoch, C., Denisova, E.A., Young, L., Qi, Z.Q., Segall-Shapiro, T.H., Calvey, C.H., Parmar, P.P., Hutchison III, C.A., Smith, H.O., Venter, J.C.: Creation of a bacterial cell controlled by a chemically synthesized genome. Science **329** (5987), 52–56 (2010)

Munteanu, A., Attolini, C.S., Rasmussen, S., Ziock, H., Solé, R.V.: Generic Darwinian selection in catalytic protocell assemblie. Philos. Trans. R. Soc. Lond. B Biol. Sci. **362**(1486), 1847–1855 (2007)

Rasmussen, S., Chen, L., Deamer, D., Krakauer, D.C., Packard, N.H., Stadler, P.F., Bedau, M. A.: Transitions from nonliving to living matter. Science **303**, 963–965 (2004)

Rasmussen, S., Bedau, M.A., Chen, L., Deamer, D., Krakauer, D.C., Packard, N.H., Stadler, P.F. (eds.): Protocells: Bridging Nonliving and Living Matter. The MIT Press, Cambridge (2008)

Serra, R., Carletti, T., Poli, I.: Synchronization phenomena in surface reaction models of protocells. Artif. Life **13**, 1–16 (2007)

Serra, R., Filisetti, A., Villani, M., Graudenzi, A., Damiani, C., Panini, T.: A stochastic model of catalytic reaction networks in protocells. Nat. Comput. **13**, 367–377 (2014)

Smith, M.J., Szathmáry, E.J.: The Major Transitions in Evolution. Oxford University Press, Oxford (1995)

Steel, M.: The emergence of a self-catalysing structure in abstract origin-of-life models. Appl. Math. Lett. **3**, 91–95 (2000)

Villani, M., Filisetti, A., Graudenzi, A., Damiani, C., Carletti, T., Serra, R.: Growth and division in a dynamic protocell model. Life **4**, 837–864 (2014)

Villani, M., Filisetti, A., Nadini, M., Serra, R.: On the dynamics of autocatalytic cycles in protocell models. In: Rossi, F., Mavelli, F., Stano, P., Caivano, D. (eds.) WIVACE 2015. CCIS, vol. 587, pp. 92–105. Springer, Cham (2016). doi:10.1007/978-3-319-32695-5_9

On the Employ of Time Series in the Numerical Treatment of Differential Equations Modeling Oscillatory Phenomena

Raffaele D'Ambrosio[1], Martina Moccaldi[1(✉)], Beatrice Paternoster[1],
and Federico Rossi[2]

[1] Department of Mathematics, University of Salerno, Fisciano, Italy
{rdambrosio,mmoccaldi,beapat}@unisa.it
[2] Department of Chemistry and Biology, University of Salerno, Fisciano, Italy
frossi@unisa.it

Abstract. The employ of an adapted numerical scheme within the integration of differential equations shows benefits in terms of accuracy and stability. In particular, we focus on differential equations modeling chemical phenomena with an oscillatory dynamics. In this work, the adaptation can be performed thanks to the information arising from existing theoretical studies and especially the observation of time series. Such information is properly merged into the exponential fitting technique, which is specially suitable to follow the a-priori known qualitative behavior of the solution. Some numerical experiments will be provided to exhibit the effectiveness of this approach.

Keywords: Oscillating solutions · Exponential fitting · Parameter estimation · Reaction equations · Belousov-Zhabotinsky reaction

1 Introduction

This work aims to solve systems of differential equations modeling oscillatory chemical phenomena. In particular, it highlights how useful can be time series of experimental data when they are properly merged into a numerical scheme.

Classic numerical methods could determine a strong reduction in stepsize in order to accurately follow the prescribed oscillations of the exact solution because they are developed in order to be exact (within round-off error) on polynomials up to a certain degree. When the qualitative behavior of the exact solution is a-priori known, it may be worthwhile to employ adapted methods which are constructed in order to be exact on functions other than polynomials, following the well-known strategy of exponential fitting [1–4]. Such functions are assumed to belong to a finite-dimensional space (the so-called fitting space) and are chosen according to the character of the exact solution. As a consequence, the coefficients of the resulting numerical method are no longer constant as in the classic case, but depend on a parameter characterizing the exact solution, whose value is evidently unknown. Therefore, the advantages of this technique

© Springer International Publishing AG 2017
F. Rossi et al. (Eds.): WIVACE 2016, CCIS 708, pp. 179–187, 2017.
DOI: 10.1007/978-3-319-57711-1_16

can be reached only if the fitting space is suitably chosen and the parameter is properly computed.

We deal with these two challenges by taking into account the existing theoretical studies on the problem and observing the time series of experimental data. The oscillatory dynamics emerging from both these approaches suggests the employ of a trigonometrical fitting space. In this case, the basis functions rely on a parameter which is the time frequency of oscillations of the exact solution. When the time series of experimental data are available, we can estimate the parameter by means of the frequency of observed oscillations, thus avoiding expensive procedures based on solving non-linear systems as in [5,6].

As an experimental case study, we focus on the Belousov-Zhabotinsky (BZ) reaction, a prototypical oscillatory chemical system whose kinetics is essentially described in the well-known *Oregonator* model developed by Field, Körös and Noyes [7–9]. It consists in a system of ordinary differential equations which we integrate by means of the above-mentioned adapted strategy.

In summary, we describe the main aspects of the Belousov-Zhabotinsky reaction in Sect. 2, Sect. 3 is devoted to the development of the numerical scheme used to integrate the Oregonator, while Sect. 4 shows some numerical experiments and Sect. 5 exhibits the conclusions.

2 The Belousov Zhabotinsky Reaction

The BZ reaction was discovered in 1951 by Boris P. Belousov who observed oscillations in the color of a solution while he was trying to develop a simple chemical model for the oxidation of organic molecules in living cells [10,11]. His study was confirmed and extended by Zhabotinsky 10 years later [12–14] and now BZ is probably one of the most studied oscillating reaction; the popularity of the BZ is mainly due to the fact that it is the simplest closed macroscopic system that can be maintained far from equilibrium by an internal source of free energy homogeneously distributed in space. Being outside of thermodynamical equilibrium, BZ displays several *exotic* dynamical regimes: periodic, aperiodic and chaotic oscillations [15,16], autocatalysis and bistability [17], Turing structures and pattern formation [18,19].

The BZ reaction consists in the oxidation of an organic substrate (generally malonic acid) by bromate ions in an acidic medium, catalyzed by a metal complex (iron, cerium or ruthenium, see [8,9] and references therein). The oscillations especially occur in the concentrations of the metal ions and become evident through a change in the color of the solution, which is more drastic for the iron. According to Fields, Körös and Noyes, the oscillations are due to the competition between two processes: firstly, the metal ion is mainly in its reduced state and the concentration of bromide ions ($[Br^-]$) is high (Process I); then the bromide ion is consumed up to a certain critical value and the metal ion reverts to the oxidized state (Process II); finally the metal ion reacts to produce bromide ions and changes to its reduced state again. However, from the kinetics point of view, oscillations are due to an Hopf instability arising from the nonlinear chemical mechanism (autocatalysis + inhibition), involved in the reaction. The whole chemical kinetics has been described by Field, Körös and Noyes by means of the following key reactions

$$A + Y \xrightarrow{k_1} X + P,$$

$$X + Y \xrightarrow{k_2} 2P,$$

$$A + X \xrightarrow{k_3} 2X + 2Z,$$

$$2X \xrightarrow{k_4} A + P,$$

$$B + Z \xrightarrow{k_5} \frac{1}{2} f Y,$$

where

$X = \text{HBrO}_2$ (bromous acid), $P = \text{HOBr}$ (hypobromous acid),
$Y = \text{Br}^-$ (bromide ion), $A = \text{BrO}_3^-$ (bromate ion),
$Z = \text{Me}^{(n+1)+}$ (metal ion in oxidized state), $B = \text{MA}$ (malonic acid).

Applying the law of mass action, the Field-Körös-Noyes model can be converted into the following third order system of kinetic equations [8]:

$$\frac{dx^*}{dt^*} = k_1 \, a \, y^* - k_2 \, x^* y^* + k_3 \, a \, x^* - 2k_4 (x^*)^2, \tag{1a}$$

$$\frac{dy^*}{dt^*} = -k_1 \, a \, y^* - k_2 \, x^* y^* + \frac{f}{2} \, k_5 \, b \, z^*, \tag{1b}$$

$$\frac{dz^*}{dt^*} = 2k_3 \, a \, x^* - k_5 \, b \, z^*, \tag{1c}$$

which is known as *Oregonator* and involve the concentrations of the aforementioned chemical elements. Such concentrations are indicated by letters in lower case henceforth. The occurrence of oscillations in the exact solution depends strongly on the values of the involved parameters, especially k_5 and f. Indeed, if $k_5 = 0$, the bromide ion (Br^-) concentration decays to zero according to the Eq. (1b), so the system cannot oscillate. With regards to f, oscillations arise only if $0.5 < f < 2.414$, whereas for $f < 0.5$ and $f > 2.414$ the reaction is in a stable steady state, being Process II or Process I dominant, respectively (see [9] and references therein).

It is more convenient to study the Oregonator (1) in its dimensionless form, as follows:

$$\epsilon \frac{dx}{dt} = q \, y - x \, y + x \, (1 - x), \tag{2a}$$

$$\epsilon' \frac{dy}{dt} = -q \, y - x \, y + f \, z, \tag{2b}$$

$$\frac{dz}{dt} = x - z, \tag{2c}$$

where

$$x = \frac{2k_4}{k_3 a} x^*, \quad y = \frac{k_2}{k_3 a} y^*, \quad z = \frac{k_4 k_5 b}{(k_3 a)^2} z^*, \quad t = \frac{t^*}{k_5 b},$$

$$\epsilon = \frac{k_5 b}{k_3 a}, \quad \epsilon' = \frac{2k_4 k_5 b}{k_2 k_3 a}, \quad q = \frac{2k_1 k_4}{k_2 k_3},$$

(3)

or, in a more compact form,

$$\frac{dr}{dt} = F(r; q, f, \epsilon, \epsilon'), \tag{4}$$

where $r = [x, y, z]^T$ and $F(r; q, f, \epsilon, \epsilon') = \begin{bmatrix} \frac{1}{\epsilon}(q\,y - x\,y + x\,(1-x)) \\ \frac{1}{\epsilon'}(-q\,y - x\,y + f\,z) \\ x - z \end{bmatrix}$.

3   An Adapted Numerical Scheme

We aim to integrate the system (4) in a certain interval $[t_0, T]$ provided with the following initial condition

$$r(t_0) = r_0, \tag{5}$$

in a region of the plane $k_5 - f$ where the solution is known to oscillate. For this purpose, we discretize the interval $[t_0, T]$ and we employ an adapted Runge Kutta method, developed in order to be exact (within round-off error) on functions belonging to a particular fitting space. The general expression of a s-stage Runge-Kutta method applied to the system (4) is

$$R_i = r_n + k \sum_{j=1}^{s} a_{i,j} F(t_n + c_j k, R_j), \quad i = 1, \ldots, s,$$

$$r_{n+1} = r_n + k \sum_{i=1}^{s} b_i F(t_n + c_i k, R_i),$$

(6)

where k is the stepsize. We remark that the system (4) is autonomous, so $F(t_n + c_j k, R_j) = F(R_j)$. The scheme (6) is a one-step procedure and each of its stages can be seen as a linear multistep formula on a non-equidistant grid [20]. Following this approach, it can be reformulated by means of the following $s+1$ linear stage representation

$$r_{n+c_i} = r_n + k \sum_{j=1}^{s} a_{i,j} F(r_{n+c_j}), \quad i = 1, \ldots, s, \tag{7a}$$

$$r_{n+1} = r_n + k \sum_{i=1}^{s} b_i F(r_{n+c_i}), \tag{7b}$$

being (7a) the internal stages and (7b) the final one. In this way, it is possible to associate a linear difference operator with each stage

$$\mathcal{L}_i[\phi(t);k] = \phi(t+c_ik) - \phi(t) - k\sum_{j=1}^{s} a_{i,j}\,\phi'(t+c_jk), \quad i = 1,\ldots,s, \qquad (8a)$$

$$\mathcal{L}[\phi(t);k] = \phi(t+k) - \phi(t) - k\sum_{i=1}^{s} b_i\,\phi'(t+k). \qquad (8b)$$

Annhilating it on a proper fitting space, we can obtain the required adapted Runge Kutta.

The prescribed oscillatory behavior of the exact solution of (4) suggests the employ of a trigonometrical fitting space

$$\mathcal{F}_{trig} = \{1, \sin(\mu t), \cos(\mu t)\}, \qquad (9)$$

and the above procedure leads to a trigonometrically fitted 2-stage Runge Kutta method having the following coefficients [21]:

$$a_{i1}(z) = \frac{1}{zD(z)} \left(\sin(c_iz)\sin(c_2z) - \cos(c_2z)(1-\cos(c_iz))\right), \ i = 1,2,$$

$$a_{i2}(z) = \frac{1}{zD(z)} \left(-\sin(c_iz)\sin(c_1z) + \cos(c_1z)(1-\cos(c_iz))\right), \ i = 1,2,$$

$$b_1(z) = \frac{1}{zD(z)} \left(\sin(z)\sin(c_2z) - \cos(c_2z)(1-\cos(z))\right),$$

$$b_2(z) = \frac{1}{zD(z)} \left(-\sin(z)\sin(c_1z) + \cos(c_1z)(1-\cos(z))\right),$$

(10)

where $z = \mu k$ and

$$D(z) = \cos(c_1z)\sin(c_2z) - \sin(c_1z)\cos(c_2z).$$

We remark that the coefficients (10) rely on the parameter μ, which needs to be properly estimated. For this purpose, we consider the experiment in [22] carried out on an unstirred ferroin catalyzed BZ system and we observe the corresponding time series reported in Fig. 1. We focus on the oscillations occurring in the concentration of ferriin, i.e. the oxidized form of the catalyst, $Fe(phen)_3^{3+}$, which corresponds to z in the *Oregonator* model (4). The time series exhibits an initial exponential decay trend corresponding to the start of the reaction. We extract the frequency of the oscillations from the time series as the inverse of the period and we use the obtained value (0.0349) as an estimate of the parameter μ. In this way, we can reach the benefits of the exponential fitting strategy without increasing the computational cost to compute an accurate estimate of μ.

4   Numerical Experiments

We now show some numerical results arising from the integration of (4) in $[0, 185]$ provided by the initial conditions

$$x(0) = 0.0013, \quad y(0) = 0.2834, \quad z(0) = 0.1984, \qquad (11)$$

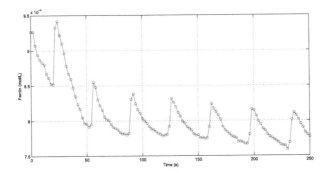

Fig. 1. Time series of concentration of ferriin related to the experiment carried out in [22] on an unstirred ferroin catalyzed BZ system.

and with the following values for the parameters

$$f = 1, \quad q = 3.52 \cdot 10^{-5}, \quad \epsilon = 0.3779, \quad \epsilon' = 7.56 \cdot 10^{-4}. \tag{12}$$

We remark that the concentrations in (11) are in their dimensionless form. We employ the trigonometrically fitted Runge Kutta method (10) described in Sect. 3 with the vector of nodes of the implicit trapezoidal rule ($c = [0, 1]$) and compare it with the corresponding classic Runge Kutta, the Lobatto-IIIA method and the well-known Gauss' Runge Kutta of order 4, which have the following Butcher's arrays [23]

$$
\begin{array}{c|cc}
0 \\
1 & 1/2 & 1/2 \\
\hline
 & 1/2 & 1/2
\end{array}
\qquad
\begin{array}{c|ccc}
0 \\
1/2 & 5/24 & 1/3 & -1/24 \\
1 & 1/6 & 2/3 & 1/6 \\
\hline
 & 1/6 & 2/3 & 1/6
\end{array}
\qquad
\begin{array}{c|cc}
1/2 - \sqrt{3}/6 & 1/4 & 1/4 - \sqrt{3}/6 \\
1/2 + \sqrt{3}/6 & 1/4 + \sqrt{3}/6 & 1/4 \\
\hline
 & 1/2 & 1/2
\end{array}
$$

respectively. Table 1 shows that the trigonometrically fitted Runge Kutta method (10) is more accurate and even stabler than the classic methods. In this table, we consider the relative error with respect to a reference solution, computed by the Matlab routine ode15s with an accuracy equal to 10^{-13}. As reported in Fig. 2, the trigonometrically fitted Runge Kutta method (10) follows the oscillations of the solution expected both from theoretical studies [9] and from the observation of time series related to the experiment in [22]. Moreover, Fig. 2 shows that the numerical solution obtained by this method and the reference solution computed by the Matlab solver ode15s exhibit totally similar oscillatory profiles. We remark that the variables concentration of ferriin (z) and time (t) have been recasted according to the positions (3).

Table 1. Comparison among some classic Runge Kutta methods and the trigonometrically fitted Runge Kutta (10) with nodes $c = [0,1]$ for the integration of system (4) with initial condition (5) and parameters chosen as in (12).

	Error		
	$k = 0.25$	$k = 1$	$k = 1.5$
Trapezoidal rule	0.000109	0.750955	0.978572
LobattoIIIA Runge Kutta	0.003356	it blows up	it blows up
Gauss' Runge Kutta	0.002093	1.000013	0.995994
Trigonometrically fitted Runge Kutta	0.000070	0.577778	0.588392

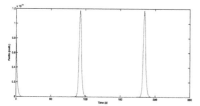 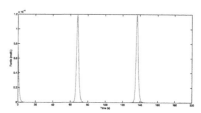

Fig. 2. Numerical solution of (4) obtained by the trigonometrically fitted Runge Kutta method (10) with nodes $c = [0,1]$ and stepsize $k = 0.25$ (on the left) and numerical solution computed by the Matlab routine **ode15s** with an accuracy equal to 10^{-13} (on the right). The variables concentration of ferriin (z) and time (t) have been recasted according to the positions (3).

5   Conclusions

In this work, we have presented an adapted numerical scheme to integrate systems of ordinary differential equations modeling oscillatory chemical phenomena. In particular, we have employed a trigonometrically fitted Runge Kutta method in order to accurately follow the prescribed oscillations of the exact solution. Indeed, such methods are constructed in order to be exact (within round-off error) on trigonometric functions. However, this strategy usually requires a further computational effort to estimate the parameter which the basis functions depend on. For this reason, we have considered the time series coming from an experiment which can be reasonably modelled by the system we want to integrate. Therefore, we have chosen the frequency of the oscillations observed in time series as an estimate of the parameter, thus avoiding an increase of computational cost. Numerical experiments show the effectiveness of this approach.

References

1. D'Ambrosio, R., Paternoster, B.: Numerical solution of reaction-diffusion systems of λ - ω type by trigonometrically fitted methods. J. Comput. Appl. Math. **294**, 436–445 (2016)
2. D'Ambrosio, R., Esposito, E., Paternoster, B.: Exponentially fitted two-step hybrid methods for $y'' = f(x, y)$. J. Comput. Appl. Math. **235**(16), 4888–4897 (2011)
3. Ixaru, L.G., Berghe, G.V.: Exponential Fitting. Springer, Netherlands (2004)
4. Paternoster, B.: Present state-of-the-art in exponential fitting. A contribution dedicated to Liviu Ixaru on his 70th birthday. Comput. Phys. Commun. **183**, 2499–2512 (2012)
5. D'Ambrosio, R., Esposito, E., Paternoster, B.: Parameter estimation in exponentially fitted hybrid methods for second order differential problems. J. Math. Chem. **50**(1), 155–168 (2012)
6. D'Ambrosio, R., Esposito, E., Paternoster, B.: Exponentially fitted two-step Runge-Kutta methods: construction and parameter selection. Appl. Math. Comp. **218**(14), 7468–7480 (2012)
7. Epstein, I.R., Pojman, J.A.: An Introduction to Nonlinear Chemical Dynamics: Oscillations, Waves, Patterns, and Chaos, 1st edn. Oxford University Press, Oxford (1998)
8. Murray, J.D.: Mathematical Biology. Springer, New York (2004)
9. Tyson, J.J.: What everyone should know about the Belousov-Zhabotinsky reaction. In: Levin, S.A. (ed.) Frontiers in Mathematical Biology. Lecture Notes in Biomathematics, vol. 100, pp. 569–587. Springer, Heidelberg (1994). doi:10.1007/978-3-642-50124-1_33
10. Belousov, B.P.: An oscillating reaction and its mechanism. In: Sborn. referat. radiat. med. (Collection of abstracts on radiation medicine), p. 145. Medgiz (1959)
11. Field, R.J., Burger, M.: Oscillations and Traveling Waves in Chemical Systems. Wiley-Interscience, New York (1985)
12. Zhabotinsky, A.M.: Periodic processes of the oxidation of malonic acid in solution (study of the kinetics of Belousov reaction). Biofizika **9**, 306–311 (1964)
13. Zaikin, A.N., Zhabotinsky, A.M.: Concentration wave propagation in two-dimensional liquid-phase self-oscillating system. Nature **225**(5232), 535–537 (1970)
14. Zhabotinsky, A.M., Rossi, F.: A brief tale on how chemical oscillations became popular: an interview with Anatol Zhabotinsky. Int. J. Des. Nat. Ecodyn. **1**(4), 323–326 (2006)
15. Marchettini, N., Budroni, M.A., Rossi, F., Masia, M., Liveri, M.L.T., Rustici, M.: Role of the reagents consumption in the chaotic dynamics of the Belousov-Zhabotinsky oscillator in closed unstirred reactors. Phys. Chem. Chem. Phys. **12**(36), 11062–11069 (2010)
16. Rossi, F., Budroni, M.A., Marchettini, N., Carballido-Landeira, J.: Segmented waves in a reaction-diffusion-convection system. Chaos: Interdisc. J. Nonlinear Sci. **22**(3), 037109 (2012)
17. Taylor, A.F.: Mechanism and phenomenology of an oscillating chemical reaction. Prog. React. Kinet. Mech. **27**(4), 247–325 (2002)
18. Budroni, M.A., Rossi, F.: A novel mechanism for in situ nucleation of spirals controlled by the interplay between phase fronts and reaction-diffusion waves in an oscillatory medium. J. Phys. Chem. C **119**(17), 9411–9417 (2015)
19. Rossi, F., Ristori, S., Rustici, M., Marchettini, N., Tiezzi, E.: Dynamics of pattern formation in biomimetic systems. J. Theor. Biol. **255**(4), 404–412 (2008)

20. Albrecht, P.: A new theoretical approach to RK methods. SIAM J. Numer. Anal. **24**(2), 391–406 (1987)
21. Paternoster, B.: Runge-Kutta(-Nyström) methods for ODEs with periodic solutions based on trigonometric polynomials. Appl. Numer. Math. **28**(2), 401–412 (1998)
22. Rossi, F., Budroni, M.A., Marchettini, N., Cutietta, L., Rustici, M., Liveri, M.L.T.: Chaotic dynamics in an unstirred ferroin catalyzed Belousov-Zhabotinsky reaction. Chem. Phys. Lett. **480**(4), 322–326 (2009)
23. Hairer, E., Lubich, C., Wanner, G.: Geometric Numerical Integration. Springer Series in Computational Mathematics, vol. 31. Springer, Heidelberg (2006)

A Program for the Solution of Chemical Equilibria Among Multiple Phases

Fulvio Ciriaco[1(✉)], Massimo Trotta[2], and Francesco Milano[2]

[1] Dip. di Chimica, Università degli studi di Bari, via Orabona 4, 70126 Bari, Italy
fulvio.ciriaco@uniba.it
[2] CNR IPCF-Bari, UOS Bari, via Orabona 4, 70126 Bari, Italy

Abstract. We present a program for the calculation of concentrations at chemical equilibrium in systems with one or more phases. We explain the main difficulties that such a program must surmount and the strategies that were devised for the present one, comparing them to others that can be found in the literature.

1 Introduction

Except for a small number of simple cases for which the computation of the equilibrium conditions of a system of reactants has precise or approximate analytical solutions, the set of nonlinear equations and inequalities that translates this general problem of chemistry can only be solved numerically.

Rather than solving the general problem, a manually simplified equation system is often written ad hoc and solved with the aid of clever choice of the variables and approximations.

Considering the importance and ubiquity of this problem, the amount of literature [1–7] and computer programs devoted to it [8–11] is outstandingly poor, perhaps because the difficulties of the solution are erroneously underestimated. Moreover, some of these programs have important limitations, being devoted to important specific cases, like combustion in the gas phase. However one can also find lots of dispersed material, mostly matlab routines using the library minimization or equation solver algorithms on a set of chemical equations that must be manually written for each separate case.

In the following sections we describe why the choice of the variables is problematic and important and the advantages of the present choice compared to the alternatives.

The precipitation equilibria are usually translated into a set of inequalities. We therefore also describe how we coped with the problem of precipitation so as to deal only with equalities.

Finally we explain the structure of the present computer library and program and the input format that was devised to describe a chemical system with sufficient generality.

© Springer International Publishing AG 2017
F. Rossi et al. (Eds.): WIVACE 2016, CCIS 708, pp. 188–197, 2017.
DOI: 10.1007/978-3-319-57711-1_17

2   The Mathematical and Thermodynamic Features of Chemical Equilibrium

We assume that the chemical system can be completely described by means of

- a number of distinct phases;
- a number of species and their propensity to transfer from a phase to another, described by transfer equations and transfer constants;
- the analytical amount for each of the species, i.e. the quantity of each that is inserted during the system preparation;
- the set of all possible reactions that the mentioned species can undergo, possibly dependant on the embedding phase; the fixed proportions characterizing the reactants disappearance and the products appearance are described by the stoichiometric coefficients and the reaction quotient at equilibrium by the equilibrium constant.

For example, the set of reactions for a water solution obtained dissolving $Fe(OH)_2$ could be schematized:

$$Fe(OH)_2 \xrightleftharpoons{K_{b1}} Fe(OH)^+ + OH^-$$

$$Fe(OH)^+ \xrightleftharpoons{K_{b2}} Fe^{2+} + OH^-$$

$$H_2O \xrightleftharpoons{K_w} H^+ + OH^-$$

$$Fe(OH)_2 \downarrow \xrightleftharpoons{K_s} Fe(OH)_2$$

where part of $Fe(OH)_2$ might be still present as a solid precipitate.

In principle one could think to obtain the solution of such a small set of relations straightforwardly writing the principle of mass conservation and the laws of thermodynamic equilibrium, e.g. for our sample system:

$$|Fe(OH)_2|_0 = |Fe(OH)_2| + |Fe(OH)^+| + |Fe^{2+}|$$
$$2|Fe(OH)_2|_0 + |H^+| = |OH^-|$$
$$\frac{[Fe(OH)^+][OH^-]}{[Fe(OH)_2]} = K_{b1}$$
$$\frac{[Fe^{2+}][OH^-]}{[Fe(OH)^+]} = K_{b2}$$
$$[H^+][OH^-] = K_w$$
$$[Fe(OH)_2] \leq K_s$$

where we indicated the concentrations in [] as usual in chemistry and the total amounts in ||. Also, it would be necessary to introduce a positiveness condition for each of the species, otherwise risking to obtain non physical solutions.

Already in this straightforward form, one has to take an important decision about the nature and number of the variables involved in the calculation. In fact, the concentration are linearly related through the mass conservation equations and these relations can be inverted to lessen the number of variables, for example this is the route taken by most programs that solve the related problem of Gibbs function minimization [3–5]. Rather than letting a linear algebra system solve the system arbitrarily, there is an intuitive set of alternative variables one can use, the advancement degree of each reaction, that is how much each of the reactions progressed relatively to the initial composition, in an arbitrary scale, usually corresponding to the amount of one of the reactants. These two positions are most often chosen in the development of chemical equilibrium software.

However, there is a third choice, i.e. to take into consideration both the reaction progress indices and the species concentrations, retaining of course the equations relating them. In fact, this was our choice, except that we preferred to retain the logarithm of the concentrations rather than the concentrations themselves.

Working with the logarithm of concentrations has three advantages:

- the concentration is bound to be positive;
- the variation of the logarithm of concentrations is similar for different species, i.e. the variation scale is homogeneous, the importance of this will be made clearer below;
- the equilibrium conditions can be easily rewritten in a general linear form: $\sum_i \chi_i ln(c_i) = ln(K)$, where χ_i are the stoichiometric coefficients, with negative sign for reactants.

The redundancy of variables has a single but very important advantage: the chemical equations can be very stiff, with equilibrium constants and concentrations easily varying by several of orders of magnitude. When working with a minimal variables set, the laws of mass conservation are automatically preserved during all the numerical solution, obliging the system to move in a possibly very narrow seam. In a redundant set of variables instead, the system deviates from solution both for the equilibrium and for the mass conservation relations and can arrive to the solutions through a path that is otherwise not accessible. The increase of the number of variables is secondary to the gained flexibility.

We also wanted to substitute the inequalities of the precipitation equilibria with equalities. The obvious explanation of the precipitation equilibrium is that the chemical potential of the precipitate is constant and factorizes out in the equilibrium law until the precipitate disappears. Our idea consists in setting the chemical potential to a function that is constant only when the formal concentration is greater than a very low threshold value, otherwise gradually resembling the common log(c) shape, as in left pane of Fig. 1. In this way, below saturation, we still have a positive amount of the precipitate but with a formal concentration so much lower than the dissociated partners that it does not significantly alter the relationship of the solute concentration to the total amount of the substance, as shown in right pane of Fig. 1.

A useful trick to emulate ideal chemical buffer, i.e. to constrain the concentration of a chemical species to a constant, is to introduce a fake precipitate source for the species, with the solubility constant equal to the required species

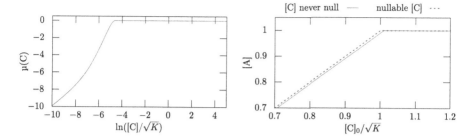

Fig. 1. On the left: our formulation of chemical potential of the C specie for the reaction $A + B \xrightleftharpoons{K} C \downarrow$, threshold value set at $\sqrt{K}/100$. On the right [A] concentrations as a function of $[C]_0$, the analytical C concentration for the classical model where [C] becomes zero under the solubility and for the model implemented in our program.

concentration. This trick also provides the amount of the chemical species that is exogenous, i.e. how much of the precipitate source was consumed.

3 The Numerical Solution

Once the set of equations is provided, the important decision must be taken of which numerical solver among many [12] to adopt.

The solution of chemical equations has specialized features that deserve at least adaptation of the available solvers. As we mentioned above, the stiffness of the chemical equation is a source of convergence and instability problems. Continuation methods are methods in which the set of equations is parameterized, usually with a single parameter that can bring the system from an initial model state, easier to solve, to the final seeked one in small steps. These methods are seldom used because the parameterization of the equations is not general and easily abstractable, leading therefore to the necessity of ad hoc solutions [13].

However, there is a very simple parameterization of the chemical equations, that can also find a thermodynamic interpretation: one can write the equilibrium constants in the form $K = e^{-\Delta G_0/RT} = e^{-\beta G_0}$, and introduce into the system a fictitious inverse temperature β. When $\beta = 0$, all the equilibrium constants are unitary and the chemical equations are much easier to solve. One can then guide the system to the final K values by small β increments.

At each of the β increments the refinement of the equilibrium is obtained by means of Newton-Raphson iterations. We opted for writing the Newton-Raphson subroutine from scratch, rather than taking one of the several implementations available, e.g. minpack. Our Newton-Raphson implementation is somewhat restricted in scope compared to others generally available, for example it requires a subroutine computing the equations jacobian whereas the minpack implementation will compute the jacobian numerically when such subroutines are not provided, however in the case of the chemical equations one should provide the jacobian, which is analytically derivable, at any rate, since its numerical computation is inefficient and subject to pitfalls. Moreover, contrarily to the

Newton-Raphson implementations of our knowledge, our own also requires a subroutine that evaluates the limit values for the variables. This dynamic estimate of the limits of the parameters is available for chemical systems and should not be neglected [2]. One of the reasons is rescaling the parameters so that they attain a metric as homogeneous as possible. This is already so for the log(c) parameters but not necessarily for the reaction advancement indices.

4 Program Description and Sample Usage

The program is splitted in two parts: a library for solution of chemical equilibrium problems, which can therefore be called directly for sofisticated usages, and a frontend program capable of parsing an input format for the description of chemical systems of reactants.

The following sample input represents the problem of benzoic acid acidity and partition between water and butanol [14] as represented by the following chemical equations:

$$BzOH(water) \xrightleftharpoons{K_{aw}} BzO^-(water) + H^+(water)$$

$$BzOH(butanol) \xrightleftharpoons{K_{ab}} BzO^-(butanol) + H^+(butanol)$$

$$BzOH(water) \xrightleftharpoons{K_p} BzOH(butanol)$$

$$H_2O(water) \xrightleftharpoons{K_w} H^+(water) + OH^-(water)$$

```
---example 1----------------------------------------------------
phases
  water
  butanol
end phases

species
H2O 55.0
H+
OH-
BzOH
BzO-
end species

equilibria
1e-14 1 H+ 1 OH- -1 H2O
6.28e-5 1 H+ 1 BzO- -1 BzOH
7.94e-16 1 butanol%H+ 1 BzO- -1 BzOH
79.6 1 butanol%BzOH -1 water%BzOH
end equilibria
```

```
composition
   water 1
      H2O 1
   end water
   butanol 1
      BzOH 0.3 0.6
   end butanol
end composition
```

There are four sections: phases, species, equilibria and composition. One can write comments outside of the sections, for example the sequence of hyphens in the header will be considered a comment.

The rules for phases are:

- each row is the name of a phase, no spaces are allowed;
- more phase names can appear than represented in the actual composition;
- the first phase name becomes the default value.

The rules for species are:

- each row is the name of a species, no spaces are allowed;
- the name may be followed by a number, representing the standard concentration of the species, as for water in water above;
- the species name can be prefixed with the phase name, e.g. "water%H2O" would specify water in water;
- there can be more species than actually appearing in the composition.

The rules for equilibria are:

- each equilibrium starts with specification of the equilibrium constant, followed by alternatively stoichiometric coefficients and species names, with reactants and products being described respectively by negative and positive coefficients;
- If a phase is mentioned, that phase becomes the default value for the current row. In example 2, the second equilibrium is e.g. all pertaining to the detergent phase.

The rules for composition are:

- the composition is a sequence of subsections describing each phase.
- the description of each phase starts with the phase name followed by its formal amount, usually the volume.
- only the analytical composition of the phase is usually needed and the program should be able to deduce all the species reachable by means of chemical equilibria. For example, from the presence of butanol%BzOH the program will allocate immediately butanol%BzO$^-$, butanol%H$^+$ and water%BzOH and subsequently water%BzO$^-$. However, the program does not make any assumption about what the phase name water means, and the species H_2O must be explicitly added to it.

– there may be a second analytical concentration in each species input, in this
  case the program will perform a ramp from one concentration to the other,
  providing results at regular intervals.

We illustrate the features of this apparently simple system solving it in the
total BzOH concentration range $10^{-7} \div 1$ M comparing BzOH, BzO$^-$ and H$^+$
when butanol is present, as in the input above, and when it is absent. The results
are illustrated in Fig. 2.

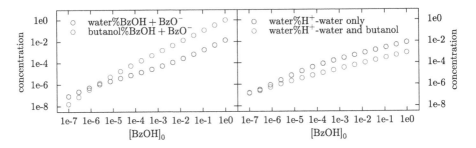

Fig. 2. In the left pane: effects of dissociation on partition of BzOH between water and
butanol. In the right pane hydrogen ion concentration for BzOH in water alone and
for BzOH in a water/butanol system.

The following points can be evidentiated:

– the dependence of pH on acid concentration is the well known sigmoid curve
  for the single phase case, however the dependence is smoother and featureless
  in the presence of butanol;
– the total concentration of BzOH is higher in water at low concentration,
  because the acidity constant and hence dissociation is much higher in water
  than in butanol. At concentrations higher than about 10^{-5} M, with dissociation decreasing, BzOH concentration is increasingly higher in the organic
  phase, in the limit of very high concentration reaching the proportions pre-
  dictable for the undissociated form alone from the partition constant.

The second sample input is for the calculation of the association equilibrium
of a small ligand to a membrane protein in a non-physiological environment. We
choose the ligand ubiquinone (UQ) which associates with a photoenzyme called
photosynthetic reaction center (RC) extracted from the bacterium Rhodobacter
sphaeroides and partitions between water and the detergent (LDAO) required to
maintain the photoenzyme in solution. The complete description of this problem
is outside the scope of this paper and can be found in [15]. The problem could
be described with the two simultaneous equilibria:

$$UQ(aq) \xrightleftharpoons{K_{rip}} UQ(det)$$

$$UQ(det) + RC(det) \xrightleftharpoons{K_b} RCUQ(det)$$

where aq and det denote the water and detergent phases respectively and RCUQ is UQ bound to RC and the first equation refers to ideal partition of UQ between water and detergent. However, the sample input refers to the similar system:

$$UQ(aq) + S(det) \xrightleftharpoons{K'_{rip}} SUQ(det)$$

$$SUQ(det) + RC(det) \xrightleftharpoons{K_b} RCUQ(det) + S(det)$$

where the solubility of UQ in detergent micelles is limited [15], as by a site or receptor model, with maximal concentration determined by the fictitious binding species S and $K'_{rip} = K_{rip}/[S]_0$ yields the same partition behaviour as the upper equations in the limit of low UQ concentration.

The amount of the detergent phase is its volume, obtained from the molar volume of the pure detergent (0.2557 L/mol) and the detergent concentration (0.39 mM).

It is customary in the literature to report the binding constant of UQ to RC referring it to the medium or bulk concentration of UQ, i.e. (UQ amount)/(total volume) rather than to the unknown UQ concentration in detergent, as if water and detergent constituted a single phase.

Such apparent binding constants can vary widely for different ubiquinones, increasing steadily with UQ hydrophobicity, as measured by K_{rip}. However a survey of the relevant literature [15–18] shows that, if the K_b is computed with reference to the UQ concentration in the micellar phase, the K_b value for different quinones is remarkably similar, about $200\,M^{-1}$.

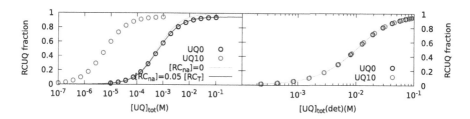

Fig. 3. The fraction of bound RC as a function of total UQ, on the left, and of total UQ in the detergent phase, on the right. The continuous curves are single phase exact solutions for the association of UQ_0 to RC.

The so called "titration curve", reported in Fig. 3, is obtained computing the chemical equilibrium for different values of [UQ]. The computed values for UQ_0 and UQ_{10} differ only in the water/detergent partition constant, respectively 8 and 10^3, assumed equal to those for water/octanol. The value 10^3 is really only a representative value for a partition constant so high that it cannot be measured.

From the computation, the "true" binding constant, referring to the UQ concentration inside the detergent phase and the "apparent" binding constant referring to the UQ concentration in the water phase can be evaluated directly.

The association of UQ to RC in a single phase has an analytical solution. This is represented by the cyan curves in Fig. 3. Of course it provides the exact ratio between the associated RC fraction and total UQ concentration in detergent, as represented in the right sub-figure. It would also represent the exact dependence against the total bulk UQ concentration for an ideal partition equilibrium, of course with scaled constants, since in that case [UQ] (water) and [UQ] (detergent) are proportional to one another. In this case, with increasing UQ amounts, the association ratio should approach unity, as depicted by the cyan line on the left of Fig. 3. The computation allows to reconduce the deviation from this behaviour not to an inaccessible RC fraction, introduced ad hoc to explain experimental results and also fitting the numerical results very well (black line of Fig. 3), but to limited UQ solubility in the detergent phase.

```
---example 2-----------------------------------------------------

phases
  water
  detergent
end phases

species
RC
RCUQ
UQ
S
SUQ
end species

equilibria
79.4 -1 detergent%S -1 water%UQ 1 detergent%SUQ
200 1 detergent%RCUQ -1 RC -1 SUQ
end equilibria

composition
  water 1
    UQ  7.2e-5
  end water
  detergent 1e-4
    RC 1e-2
   S 0.1
end detergent
end composition
```

5   Conclusions

The solution of chemical equilibria for many species or phases systems is not straightforward. It took some effort to write a program that both suites our computational needs and accepts an intuitive and flexible input. We decided to share this effort; the program will therefore be made opensource, under the name ChemEq, as soon as this description, which is also meant to document the program becomes public.

References

1. Greiner, H.: An efficient implementation of Newtons's method for complex chemical equilibria. Comput. Chem. Eng. **15**, 115–123 (1991)
2. Carrayrou, J., Mosé, R., Behra, P.: New efficient algorithm for solving thermodynamic chemistry. AIChE J. **48**, 894–904 (2002)
3. Paz-García, J.M., Johannesson, B., Ottosen, L.M., Ribeiro, A.B., Rodríguez-Maroto, J.M.: Computing multi-species chemical equilibrium with an algorithm based on reaction extents. Comput. Chem. Eng. **58**, 135–143 (2013)
4. Koukkari, P., Pajarre, R.: A Gibbs energy minimization method for constrained and partial equilibria. Pure Appl. Chem. **83**, 1243–1254 (2011)
5. Néron, A., Lantagne, G., Marcos, B.: Computation of complex and constrained equilibria by minimization of the Gibbs free energy. Chem. Eng. Sci. **82**, 260–271 (2012)
6. Kirkner, D.J., Reeves, H.W.: A penalty function method for computing chemical equilibria. Comput. Geosci. **16**, 21–40 (1990)
7. Meintjes, K., Morgan, A.P.: A methodology for solving chemical equilibrium systems. Appl. Math. Comput. **22**, 333–361 (1987)
8. NASA: Chemical Equilibrium with Applications. http://www.grc.nasa.gov/WWW/CEAWeb/. freeware
9. Mathtrek Systems: Eqs4win. http://www.mathtrek.com/
10. GTT-Technologies: Chemsage. http://gtt.mch.rwth-aachen.de/gtt-web/chemsage
11. OLI Systems Inc.: Oli Aqueous Electrolyte Models. http://www.olisystems.com/
12. Dent, D., Paprzycki, M., Kucaba-Piętal, A.: Comparing solvers for large systems of nonlinear algebraic equations. In: Proceedings of the Southern Conference on Computing the University of Southern Mississippi, 26–28 October 2000
13. Rheinboldt, W.: Numerical Analysis of Parameterized Nonlinear Equations. Wiley, New York (1986). ISBN: 0-471-88814-1
14. Sarmini, K., Kenndler, E.: Ionization constants of weak acids and bases in organic solvents. J. Biochem. Biophys. Methods **38**, 123–137 (1999)
15. Ciriaco, F., Tangorra, R.R., Antonucci, A., Giotta, L., Agostiano, A., Trotta, M., Milano, F.: Semiquinone oscillations as a tool for investigating the ubiquinone binding to photosynthetic reaction centers. Eur. Biophys. J. **44**, 183–192 (2015)
16. Shinkarev, V.P., Wraight, C.A.: The interaction of quinone and detergent with reaction centers of purple bacteria. I. Slow quinone exchange between reaction center micelles and pure detergent micelles. Biophys. J. **72**, 2304–2319 (1997)
17. McComb, J.C., Stein, R.R., Wraight, C.A.: Investigations on the influence of head-group substitution and isoprene side-chain length in the function of primary and secondary quinones of bacterial reaction centers. Biochimica et biophysica acta **1015**, 156–171 (1990)
18. Mallardi, A., Palazzo, G., Venturoli, G.: Binding of ubiquinone to photosynthetic reaction centers: determination of enthalpy and entropy changes in reverse micelles. J. Phys. Chem. B **101**, 7850–7857 (1997)

Author Index

Amodio, Michele   65
Amoretti, Michele   14

Bellantuono, Giuseppe   65
Bevilacqua, Vitoantonio   65
Braccini, Michele   91
Brunetti, Antonio   65
Bruni, Martino   65
Budroni, Marcello A.   3
Buonamassa, Giuseppe   65

Cagnoni, Stefano   14
Caldarelli, Guido   26
Calvanese, Giordano   167
Cattaneo, Giuseppe   53
Ciriaco, Fulvio   188
Colombo, Riccardo   126, 138

D'Ambrosio, Raffaele   179
Damiani, Chiara   126, 138
Delfine, Giancarlo   65
Di Biasi, Luigi   53
Di Filippo, Marzia   126, 138
Di Nardo, Armando   26

Emmert-Streib, Frank   161

Facchini, Angelo   26
Ferraro Petrillo, Umberto   53, 77

Gentili, Pier Luigi   151
Giancarlo, Raffaele   53
Giovannelli, Alessandro   103
Guerriero, Andrea   65

Khoroshiltseva, Marina   103, 114

Lattanzi, Nicola   26
Liberatore, Giovanni   26
Lunardon, Nicola   114

Magaletti, Domenico   65
Mameli, Valentina   114
Maso, Lorenzo Dal   26
Mauri, Giancarlo   126, 138
Milano, Francesco   188
Moccaldi, Martina   179
Mordonini, Monica   14

Pastor-Satorras, Romualdo   3
Paternoster, Beatrice   179
Pecori, Riccardo   14
Pescini, Dario   126, 138
Piotto, Stefano   53
Poli, Irene   103, 114

Riezzo, Marco   65
Righi, Riccardo   42
Roli, Andrea   14, 42, 91
Roscigno, Gianluca   53
Rossi, Federico   179
Russo, Margherita   42

Sani, Laura   14
Scala, Antonio   26
Serra, Roberto   14, 42, 91, 167
Slanzi, Debora   103, 114

Trotta, Gianpaolo Francesco   65
Trotta, Massimo   188

Verrino, Luca   65
Vicari, Emilio   14
Villani, Marco   14, 42, 91, 167
Vitali, Roberto   77
Yli-Harja, Olli   161
Yli-Hietanen, Jari   161

·

Printed in the United States
By Bookmasters